CONTEMPORARY MATHEMATICS

Robert L. Wheeler

Titles in this Series

Problems of Elastic Stability and Vibrations

CONTEMPORARY MATHEMATICS

Volume 4

Problems of Elastic Stability and Vibrations

AMERICAN MATHEMATICAL SOCIETY
Providence · Rhode Island

PROCEEDINGS OF THE SPECIAL SESSION ON
PROBLEMS IN ELASTIC VIBRATIONS, STABILITY AND RELATED TOPICS
786th MEETING OF THE AMERICAN MATHEMATICAL SOCIETY

HELD AT DUQUESNE UNIVERSITY
PITTSBURGH, PENNSYLVANIA

MAY 15–16, 1981

EDITED BY

VADIM KOMKOV

1980 Mathematics Subject Classification. 73H05, 73H10, 73D30.

Library of Congress Cataloging in Publication Data
Main entry under title:

Problems of elastic stability and vibrations.

(Contemporary mathematics; v. 4)
"Enlarged versions of the talks presented at the spring meeting (May 15–16, 1981) of the American Mathematical Society held in Pittsburgh, PA" – Introd.
Includes bibliographies.
Contents: On the stabilization of ill posed Cauchy problems in nonlinear elasticity / L. E. Payne–A nonlinear eigenvalue problem with an exponential nonlinearity / James Moseley–Eigenvalue problems for variational inequalities / Erich Miersemann–[etc.]
1. Elasticity–Addresses, essays, lectures. 2. Vibrations–Addresses, essays, lectures. I. Komkov, Vadim. II. American Mathematical Society. III. Series: Contemporary mathematics (American Mathematical Society); v. 4.
TA653.P76 1981 531′.3823 81-12833
ISBN 0-8218-5005-9 AACR2
ISSN 0271 -4132

CONTENTS

INTRODUCTION

The articles collected in this volume are enlarged versions of the talks presented at the spring meeting (May 15-16, 1981) of the American Mathematical Society held in Pittsburgh, PA. Professor Miersemann could not attend but submitted a paper for consideration in this collection.

All papers are directly related to problems in stability or eigenvalue problems arising in considerations of elastic stability or elastic vibrations. A multitude of papers have appeared in the five years preceding this meeting in various aspects of elastic stability and elastic vibrations because of the rapidly growing theoretical developments in structural and mechanical design problems.

An unexpected turn of events in the optimal design theory necessitated a reexamination of the "usual" approach to the entire area of elastic stability and vibrations.

In papers published before 1970 most authors, including J. B. Keller, in his pioneering article (The shape of the strongest column, Archive for rational mechanics and analysis, vol. 5, #4, 1960 p. 275–285) assumed that optimizing structures against buckling should produce solutions which are reasonably smooth, or at least continuous. In recent years a critique of the early papers was offered independently by several authors such as N. Olhoff, E. F. Masur, E. J. Haug, who observed that for example the predictions of J. B. Keller or I. Tadjbakhsh and Keller appear to be in error. A number of cases in structural stability and vibrations appeared to have singular solutions. In particular discontinuities were discovered in the displacement fields.

Such singularities at points of zero cross-sectional area correspond to occurrence of plastic hinges.

Many cases involving singularities were discovered to contain the occurrence of multiple eigenvalues.

Also, the designs "close to singular" appeared to be very sensitive to effects of terms which are normally neglected (because of "smallness") prompting some authors (E. Haug and V. Komkov) to re-examine the commonly used hypothesis of the practical implications to engineering design. These new research directions were reinforced by corresponding progress in purely functional analytic techniques (Zolezzi, J. Cea, E. Haug, and B. Rousselet), variational techniques (E. Miersemann, V. Komkov, J. E. Taylor) and a variety of numerical approaches to elastic stability.

Independently, some significant progress was made in the elastic stability theory by modern applications of bifurcation theory (D. Sather and G. H. Knightly). The multiple eigenvalue problem which turns up in engineering design problems has been investigated in recent papers of D. H. Sattinger, but aside from a few pioneering papers such as the one by Knightly and Sather offered in this collection, these techniques have not filtered down to the advanced level of engineering research. The much more difficult problems of stabilizability of instable elastic structures may still be considered to be practically untouched by modern mathematical techniques.

In a series of papers authored by R. J. Knops, L. E. Payne and H. H. Levine, the Cauchy problem in classical nonlinear dynamics, and in particular, the stability of the null solution was investigated.

The paper of L. E. Payne given in this collection offers some stabilizability results for certain classes of Cauchy problems in nonlinear hyperelasticity. It may be regarded as a first step towards solving a class of very difficult problems.

No meeting discussing applied problems can be regarded as a complete success unless someone presents a study combining known numerical results, experimental facts and theoretically predicted behavior. The paper of R. H. Plaut summarizes many techniques and results of engineering studies concerning the vibration and stability of shallow arches. The results given here offer an insight into behavior of other shallow elastic structures which are much more difficult to analyze. Some practical results quoted by Plaut have been confirmed experimentally, offering valuable guidelines for developing future mathematical models.

Professor J. Mikusiński offered once the following advice to applied mathematicians. Do not argue with the engineers who know what the answers should look like. Instead look for a better mathematical model if your predictions do not agree with theirs. Hopefully, our theoretical predictions concerning complex stability and vibration problems begin to look realistic.

Vadim Komkov

Contemporary Mathematics
Volume 4
1981

ON THE STABILIZATION OF ILL POSED CAUCHY PROBLEMS IN NONLINEAR ELASTICITY

L.E. Payne[1]

ABSTRACT. Little is known about continuous data dependence in the Cauchy problem of nonlinear hyperelasticity except for special choices of the strain energy function. The aim of this paper is to determine stabilizing criteria for certain classes of Cauchy problems whose solutions are inherently unstable. The results described herein are still very special, and should be regarded only as a first step toward the resolution of the more general problem.

1. INTRODUCTION. In this paper we are concerned primarily with the question of continuous dependence on the data (over a finite time interval) for solutions of the dynamical Cauchy problem of nonlinear hyperelasticity, but we also make some remarks on the Cauchy problem in nonlinear elastostatics. In general the problems treated will be ill posed with solutions (if they exist) which are inherently unstable. Our aim is to determine constraint sets which have the property that solutions which lie in such a set are stable in the sense of Hadamard. More precisely, solutions which lie in such a constraint set will depend continuously on the data (in some appropriate measure) over a given finite time interval.

It is well known (see e.g. Knops, Levine and Payne [3], Knops [2] and the papers cited therein) that unless the strain energy function is suitably restricted no solution to the dynamical Cauchy problems of nonlinear elasticity can exist for all time. When solutions fail to exist globally one can frequently obtain an upper bound for the time of breakdown of solution. Thus if the solution must break down prior to a time t_1 and one asks for a nonzero solution which is uniformly bounded in the appropriate norm on the interval $[0,t_2]$ where $t_2 > t_1$ then one is discussing a nonexistent solution. We shall not, however, be concerned with the existence question in this paper.

We shall be dealing with a hyperelastic material whose density ρ is independent of time and strictly positive throughout the region of definition

1980 Mathematics Subject Classification 73H10, 73C50, 35R25.
[1]Supported in part by the National Science Foundation Grant NSF MCS 79-19358

$\Omega \times [0,T)$. Here Ω is a bounded region in R^3 with Lipschitz boundary $\partial\Omega$. The first two theorems impose strong restrictions on the strain energy function W and rather severe regularity assumptions on the solution vector itself. Our goal is to determine criteria which will guarantee that the displacement components $u_i(x,t)$ and $v_i(x,t)$ corresponding to two different sets of data that are "close" in some appropriate measure will remain "close" over some fixed time interval $0 \leq t \leq t^* < T$.

There are few results in the literature on the stabilizing of ill posed Cauchy problems in nonlinear elasticity. The results we present for well posed problems are known even though the specific inequalities exhibited in this paper have not, to the authors knowledge, appeared in print. The elastostatics results, which we describe briefly, were obtained by Knops and Payne [5]. Perhaps the special results exhibited here will interest others in attacking the stability question for more general classes of problems.

2. THE DYNAMICAL PROBLEM. For the equations of classical linear elasticity, Knops and Payne [4] developed inequalities which led to results on continuous dependence on various types of data even when the "potential energy" term was permitted to be nonpositive. In fact these authors established stability in the sense of Hadamard on finite time intervals under the assumption that solutions should be in some appropriate constraint set. As we indicated in the introduction, the analogous results for nonlinear hyperelasticity are meager except in the case of perturbations of the null solution.

In order to display the problems more compactly we shall use the index notation, and the summation convention (over repeated indices). We wish to compare solutions (assumed to exist) of the following two problems:

$$\rho(x)\frac{\partial^2 u_i}{\partial t^2} - \frac{\partial}{\partial x_j}\left(\frac{\partial W(\nabla u)}{\partial(\partial u_i/\partial x_j)}\right) = F_i(x,t) \quad \text{in} \quad \Omega \times (0,T)$$

$$(2.1) \qquad u_i = f_i(x,t) \quad \text{on} \quad \partial\Omega \times [0,T)$$

$$u_i(x,0) = g_i(x); \quad \frac{\partial u_i}{\partial t}(x,0) = h_i(x), \quad x \in \Omega$$

and

$$\rho^*(x)\frac{\partial^2 v_i}{\partial t^2} - \frac{\partial}{\partial x_j}\left(\frac{\partial W(\nabla v)}{\partial(\partial v_i/\partial x_j)}\right) = F_i^*(x,t) \quad \text{in} \quad \Omega \times (0,T)$$

$$(2.2) \qquad v_i = f_i(x,t) \quad \text{on} \quad \partial\Omega \times [0,T)$$

$$v_i(x,0) = g_i^*(x); \quad \frac{\partial v_i}{\partial t}(x,0) = h_i^*(x), \quad x \in \Omega.$$

Here we have used the notation $W(\nabla u)$ to indicated that W is a function of the displacement gradients. However, we shall not impose the usual form invariance requirements on W. The problem would have become considerably more complicated if we had chosen the boundary values for v_i to be different from those for u_i, but it is possible to obtain somewhat weaker results in that case.

Now set

$$\phi_i = u_i - v_i \tag{2.3}$$

and observe that in $\Omega \times (0,T)$, ϕ_i satisfies the equation

$$\rho \frac{\partial^2 \phi_i}{\partial t^2} + (\rho-\rho^*) \frac{\partial^2 v_i}{\partial t^2} - \frac{\partial}{\partial x_j}\left\{\frac{\partial W(\nabla u)}{\partial(\partial u_i/\partial x_j)} - \frac{\partial W(\nabla v)}{\partial(\partial v_i/\partial x_j)}\right\} = F_i - F_i^* \tag{2.4}$$

as well as the conditions

$$\phi_i = 0 \quad \text{on} \quad \partial\Omega \times [0,T) \tag{2.5}$$

$$\phi_i(x,0) = g_i - g_i^*;\ \frac{\partial \phi_i}{\partial t}(x,0) = h_i - h_i^*,\ x \in \Omega.$$

We wish to show that under certain hypotheses of the form of W and possible constraint set restrictions, if the data are "close" then the solutions u_i and v_i will be close in the appropriate sense. The result will be in the form of a specific bound for some norm of ϕ_i in terms of norms of the differences of data terms.

The first result which we shall mention seems to be rather well known, but the author did not find the result written down in the form in which it is presented here. In any case we shall need some of the same arguments in the second case to be considered. We now establish

THEOREM 1. <u>Let W be a C^3 function of its argument then if u_i and v_i are classical solutions of their respective Cauchy problems and W satisfies</u> the inequality

$$\frac{\partial^2 W(\nabla u)}{\partial u_{i,j}\partial u_{k,\ell}} \psi_{ij}\psi_{k\ell} \geq c_0 \psi_{ij}\psi_{ij},\ \forall \psi_{ij} (c_0 > 0) \tag{2.6}$$

<u>it follows that</u>

$$\int_\Omega [\rho \frac{\partial \phi_i}{\partial t}\frac{\partial \phi_i}{\partial t} + \phi_{i,j}\phi_{i,j}]dx \tag{2.7}$$

$$\leq k_1(t)||\nabla(g-g^*)||^2 + k_2(t)||h-h^*||^2 + k_3(t)\int_0^t ||(\rho-\rho^*)\frac{\partial^2 v}{\partial \eta^2} - (F-F^*)||^2 d\eta.$$

In (2.6) we have used a comma to indicated differentiation with respect to a space variable and in (2.7) the notation

$$||\nabla H||^2 = \int_\Omega H_{i,j}H_{i,j}dx;\ ||G||^2 = \int_\Omega G_i G_i dx. \tag{2.8}$$

The functions $k_i(t)$ are continuous functions of t on the interval $[0,T]$ and c_0 is a positive constant. It should be remarked that the definiteness condition (2.6) implies that the system in (2.1) is hyperbolic. It can be shown that this implies that (2.1) is a well posed problem.

To further simplify the notation we shall denote the time derivative by a dot, i.e.

$$\dot{u}_i \equiv \frac{\partial u_i}{\partial t}. \tag{2.9}$$

To prove the theorem we form

$$\int_0^t\int_\Omega \dot{\phi}_i\left\{\rho\ddot{\phi}_i+(\rho-\rho^*)\ddot{v}_i - \frac{\partial}{\partial x_j}\left[\frac{\partial W(\nabla u)}{\partial u_{i,j}} - \frac{\partial W(\nabla v)}{\partial v_{i,j}}\right]-(F_i-F_i^*)\right\}dx\,d\eta = 0. \tag{2.10}$$

Integration by parts and use of the mean value theorem lead to

$$\int_0^t\int_\Omega \left\{\rho\dot{\phi}_i\ddot{\phi}_i + B_{ijk\ell}\,\dot{\phi}_{i,j}\phi_{k,\ell} + [(\rho-\rho^*)\ddot{v}_i - (F_i - F_i^*)]\dot{\phi}_i\right\}dx\,d\eta = 0, \tag{2.11}$$

where

$$B_{ijk\ell} = \frac{\partial^2 W(\nabla\sigma)}{\partial u_{i,j}\partial u_{k,\ell}}, \tag{2.12}$$

the argument $\nabla\sigma$ being some value between ∇u and ∇v. We integrate the second term on the right by parts (noting that $B_{ijk\ell} = B_{k\ell ij}$) to obtain

$$\frac{1}{2}\int_\Omega[\rho\dot{\phi}_i\dot{\phi}_i + B_{ijk\ell}\phi_{i,j}\phi_{k,\ell}]dx\Big|_0^t = \frac{1}{2}\int_0^t\int_\Omega \dot{B}_{ijk\ell}\,\phi_{i,j}\phi_{k,\ell}\,dx\,d\eta \tag{2.13}$$
$$+ \int_0^t\int_\Omega [(\rho-\rho^*)\ddot{v}_i-(F_i-F_i^*)]\dot{\phi}_i\,dx\,d\eta.$$

Since W is a C^3 function of its argument we obtain (making use of (2.6))

$$E(t) \equiv \frac{1}{2}\int_\Omega[\rho\dot{\phi}_i\dot{\phi}_i + B_{ijk\ell}\phi_{i,j}\phi_{k,\ell}]dx \le E(0) + k\int_0^t\int_\Omega B_{ijk\ell}\phi_{i,j}\phi_{k,\ell}\,dx\,d\eta \tag{2.14}$$
$$+ \int_0^t\int_\Omega [(\rho-\rho^*)\ddot{v}_i + (F_i-F_i^*)]\dot{\phi}_i\,dx\,d\eta,$$

where k is a positive constant. Clearly then by using the arithmetic-geometric mean inequality on the last term we obtain for some constants, c_i

$$E(t) \le E(0) + c_1\int_0^t E(\eta)d\eta + c_2\int_0^t ||(\rho-\rho^*)\ddot{v} - (F_i - F_i^*)||^2 d\eta \tag{2.15}$$

from which it follows by integration that

$$E(t) \le E(0)e^{c_1 t} + c_2\int_0^t e^{c_1(t-\eta)}||(\rho-\rho^*)\ddot{v} + (F_i-F_i^*)||^2 d\eta \tag{2.16}$$

which is equivalent to (2.7).

Little is known about continuous dependence on the data when (2.6) is relaxed. The solution of (2.1) will then in general be inherently unstable. There is, however, one case in which it is possible to retrieve continuous dependence (in the sense of John [1]).

Suppose that instead of (2.6), the function W, again assumed to be a C^3 function of its argument, satisfies

$$B_{ijk\ell}\,\psi_{ij}\psi_{k\ell} \le -\,c_0\,\psi_{ij}\psi_{ij},\quad \forall\psi_{ij}\quad (c_0 > 0). \tag{2.17}$$

We define a constraint set M as the set of L_2 functions satisfying

$$\int_0^t\int_\Omega \rho u_i u_i\,dx\,d\eta \le m^2 \tag{2.18}$$

and establish the following theorem:

THEOREM II. <u>In the space of classical solutions of (2.1) which belong to M and for which (2.17) applies and</u>

$$B_{ijk\ell}\phi_{i,j}\phi_{k,\ell} \le K\phi_{i,j}\phi_{i,j}, \tag{2.19}$$

<u>the following inequality holds for $t \in (0,T)$:</u>

$$\int_0^t\int_\Omega \rho\phi_i\phi_i\,dx\,d\eta \le K\,Q^{2(1-\delta)}m^{2\delta}\qquad 0 < \delta < 1. \tag{2.20}$$

<u>Here δ is a computable function of t and Q is given by</u>

$$Q^2 = \gamma_1\int_\Omega \rho(g_i-g_i^*)(g_i-g_i^*)dx + \gamma_2\int_\Omega \rho(h_i-h_i^*)(h_i-h_i^*)dx + \gamma_3\int_0^T ||(\rho-\rho^*)v - (F-F^*)||^2 d\eta, \tag{2.21}$$

<u>where the γ_i are suitably chosen constants.</u>

We observe first that from (2.13) we obtain instead of (2.14),

$$E(t) \le E(0) + \alpha_1\int_0^t\int_\ell [\rho\dot\phi_i\dot\phi_i - B_{ijk\ell}\phi_{i,j}\phi_{k,\ell}]dx\,d\eta + \alpha_2\int_0^t ||(\rho-\rho^*)v-(F-F^*)||^2 d\eta, \tag{2.22}$$

where we have made use of (2.19). We shall need this inequality later.

To prove Theorem II we define

$$\Phi = \int_0^t\int_\Omega \rho\phi_i\phi_i\,dx\,d\eta + (T-t)\int_\Omega \rho(g_i-g_i^*)(g_i-g_i^*) + \alpha_3 Q^2, \tag{2.23}$$

where Q^2 is given by (2.21). Then

$$\dot\Phi = 2\int_0^t\int_\Omega \rho\phi_i\,\dot\phi_i\,dx\,d\eta \tag{2.24}$$

and

$$(2.25)\quad \ddot{\Phi} = 2\int_\Omega \rho\phi_i\dot{\phi}_i dxd\eta$$

$$= 2\int_\Omega \rho(g_i-g_i^*)(h_i-h_i^*)dx + 2\int_0^t\int_\Omega \rho\rho\dot{\phi}_i\dot{\phi}_i dxd\eta + 2\int_0^t\int_\Omega \rho\phi_i\ddot{\phi}_i dxd\eta.$$

We now insert the differential equation on the right, integrate by parts and use the mean value theorem to obtain

$$(2.26)\quad \ddot{\Phi} = 2\int_\Omega \rho(g_i g_i^*)(h_i-h_i^*)dx + 2\int_0^t\int_\Omega \rho\dot{\phi}_i\dot{\phi}_i dxd\eta - 2\int_0^t\int_\Omega B_{ijk\ell}\phi_{i,j}\phi_{k,\ell}dxd\eta$$

$$- 2\int_0^t\int_\Omega [(\rho-\rho^*)\ddot{v}_i - (F_i-F_i^*)]\phi_i dxd\eta.$$

Our aim is to show that Φ satisfies a differential inequality of the form

$$(2.27)\quad \Phi\ddot{\Phi} - (\dot{\Phi})^2 \geq -\beta_1\Phi\dot{\Phi} - \beta_2\Phi^2$$

from which (2.20) will follow. In fact it is easily seen that

$$(2.28)\quad \ddot{\Phi} \geq 2\int_0^t\int_\Omega \rho\,\dot{\phi}_i\dot{\phi}_i dx - 2\int_0^t\int_\Omega B_{ijk\ell}\phi_{i,j}\phi_{k,\ell}dxd\eta - \gamma_4\Phi.$$

Integrating this expression we obtain

$$(2.29)\quad \dot{\Phi}(t) - \dot{\Phi}(0) \geq 2\int_0^t\int_\Omega (t-\eta)[\rho\dot{\phi}_i\dot{\phi}_i - B_{ijk\ell}\phi_{i,j}\phi_{k,\ell}]dxd\eta - \gamma_4 T\Phi,$$

while an integration of (2.22) gives

$$(2.30)\quad \int_0^t E(\eta)d\eta \leq TE(0) + \alpha_1\int_0^t\int_\Omega (t-\eta)[\rho\dot{\phi}_i\dot{\phi}_i - B_{ijk\ell}\phi_{i,j}\phi_{k,\ell}]dxd\eta$$

$$+ \alpha_2\int_0^t\int_\Omega (t-\eta)||(\rho-\rho^*)\ddot{v} - (F-F^*)||^2 d\eta.$$

Combining we find

$$(2.31)\quad \int_0^t E(\eta)d\eta \leq \frac{\alpha_1}{2}[\dot{\Phi}(t) - \Phi(0)] + \frac{\gamma_4}{2}T\Phi - \alpha_2 T\int_0^t ||(\rho-\rho^*)\ddot{v}-(F-F^*)||^2 d\eta$$

which reduces to

$$(2.32)\quad \int_0^t E(\eta)d\eta \leq \frac{\alpha_1}{2}\dot{\Phi} + \alpha_3\Phi$$

or

$$(2.33)\quad \int_0^t\int_\Omega \rho\dot{\phi}_i\dot{\phi}_i\, dxd\eta + \int_0^t\int_\Omega B_{ijk\ell}\phi_{i,j}\phi_{k,\ell}dxd\eta \leq \frac{\alpha_1}{2}\dot{\Phi} + \alpha_3\Phi.$$

Eliminating the second term on the right of (2.28) by use of the expression above we obtain

(2.34) $$\ddot{\Phi} \geq 4 \int_0^t \int_\Omega \rho \dot{\phi}_i \dot{\phi}_i \, dx\, d\eta - \beta_1 \dot{\Phi} - \beta_2 \Phi,$$

which by use of Schwarz's inequality leads directly to the desired inequality (2.27). It is well known that inequality (2.27) may be integrated to give (2.20) for solutions in the set M.

If we take F_i and f_i to be zero in (2.1) and assume v_i to be the null solution, then continuous dependence under perturbations of the null solution is rather easy to derive under much milder assumptions on the form of W. In fact u_i need only be a weak solution of (2.1).

The main simplification when $v_i \equiv 0$ comes in the energy expression which results from (2.10). We now have for strong solutions

(2.35) $$E(t) \equiv \frac{1}{2} \int_\Omega \rho \dot{u}_i \dot{u}_i \, dx + \int_\Omega W \, dx = E(0).$$

In the weak setting it is customary to postulate

(2.36) $$E(t) \leq E(0).$$

If we define M_1 to be the set of weak solutions of (2.1) (with F_i and f_i identically zero) which further satisfy the inequality

(2.37) $$\int_\Omega \rho u_i(x,T) u_i(x,T) \, dx \leq m_1,$$

then we have the following:

THEOREM III: <u>Let u_i be a weak solution of (2.1) with F_i and f_i equal zero then if</u>

(2.38) $$u_{i,j} \frac{\partial W}{\partial (u_{i,j})} \leq 2W$$

<u>it follows that for $u_i \in M_1$</u>

(2.39) $$\Phi_1(t) \leq K(\Phi_1(0))^{1-t/T} \, m_1^{2t/T}.$$

<u>On the other hand if</u>

(2.40) $$u_{i,j} \frac{\partial W}{\partial (u_{i,j})} + 2W \geq 0$$

<u>it follows that</u>

(2.41) $$\Phi_1 \leq \Phi_1(0) + t\, \Phi_1'(0) + 2t^2 E(0).$$

<u>where</u>

(2.42) $$\Phi_1(t) = \int_\Omega \rho u_i u_i \, dx + c_1 E(0).$$

We present the proof for strong solutions. The corresponding result for weak solutions follows from analogous reasoning.

Both results follow directly from second order differential inequalities for Φ_1. We note that

(2.43) $$\dot{\Phi}_1(t) = 2 \int_\Omega \rho u_i \dot{u}_i \, dx$$

and

$$\ddot{\Phi}_1(t) = 2 \int_\Omega \rho\dot{u}_i\dot{u}_i dx - 2\int_\Omega u_{i,j} \frac{\partial W}{\partial u_{i,j}} dx. \tag{2.44}$$

Thus (2.38) and (2.35) lead to

$$\ddot{\Phi}_1(t) \geq 4 \int_\Omega \rho\dot{u}_i\dot{u}_i dx - 4E(0), \tag{2.45}$$

while (2.40) yields

$$\ddot{\Phi}_1(t) \leq 4E(0). \tag{2.46}$$

The result (2.41) follows from a direct integration of (2.46). On the other hand from (2.45) we obtain via Schwarz's inequality

$$\Phi_1\ddot{\Phi}_1 - (\dot{\Phi}_1)^2 \geq -4E(0)\Phi_1 \geq -\frac{4}{c_1}\Phi_1^2, \tag{2.47}$$

which integrates to give (2.39). Both (2.39) and (2.41) are continuous dependence inequalities in the appropriately defined spaces.

It is clear also from (2.35) (or (2.36)) that if W is nonnegative then the solution depends continuously on the data (in L_2) for $0 \leq t < T$. This follows since

$$\int_\Omega \rho u_i\dot{u}_i dx \leq \left\{\int_\Omega \rho u_i u_i dx \int_\Omega \rho\dot{u}_i\dot{u}_i dx\right\}^{1/2} \tag{2.48}$$

or

$$\dot{\Phi}_1 \leq 2\sqrt{2}\,\Phi_1^{1/2}[E(0)]^{1/2}. \tag{2.49}$$

An integration yields the inequality

$$\Phi_1(t) \leq [\Phi_1^{1/2}(0)4\sqrt{2}\; t\; E^{1/2}(0)]^2. \tag{2.50}$$

It is clear from the foregoing that if W is nonnegative or if (2.40) is satisfied then the null solution is stable with no a priori constraint set restrictions. On the other hand if (2.38) is satisfied and $W \not> 0$ the problem may be inherently unstable.

3. THE CAUCHY PROBLEM IN ELECTROSTATICS. Very little attention has been given to the ill posed Cauchy problem in elastostatics even in the linear case. We mention here some recent results of Knops and Payne [5] for a strain energy function W of rather special type. The problem considered is the following:

$$\begin{aligned} &\frac{\partial}{\partial x_j}\left(\frac{\partial W}{\partial u_{i,j}}\right) = 0 \quad \text{in } \Omega \\ &u_i = f_i \quad \text{on } \Sigma \\ &\frac{\partial W}{\partial u_{i,j}} n_j = g_i \quad \text{on } \Sigma, \end{aligned} \tag{3.1}$$

where Σ is a connected portion of the boundary $\partial\Omega$ of nonzero measure and

n_j is the jth component of the unit normal (directed outward from Ω).

The special choice for W which we consider is

$$W(\nabla u) \equiv W(\sigma), \tag{3.2}$$

where

$$\sigma = a_{rk} u_{i,r} u_{i,k}. \tag{3.3}$$

In what follows we use the notation

$$W'(\sigma) = 2p(\sigma). \tag{3.4}$$

The function $p(\sigma)$ is assumed to satisfy the following hypotheses:

H.1. $p(s) > 0$ for $s > 0$

H.2. $\int_0^s p(t)dt \le sp(s)$.

By introducing a suitable family of smooth surfaces parametrized by a parameter α, $(0 \le \alpha \le 1)$ it is possible to derive an inequality which implies the stability of the null solution in a subregion Ω_α for solutions in the appropriate constraint set. The constraint set employed by the authors was quite restrictive, i.e. the solution must be a classical solution in Ω which in that region satisfies

$$\sup_{x \in \Omega_1} \left| \frac{p'(\sigma)}{p(\sigma)} a_{rk} u_{i,r} u_{i,kj} \right| \le A_j ; \qquad \int_{\Omega_1} p(\sigma) u_i u_i dx \le k, \tag{3.5}$$

for constants A_j and k. The proof makes use of logarithmic convexity argument with α playing the role of t. For details the reader is referred to [5].

It should be noted that the linearized version of our problem does not correspond to linear anisotropic elasticity. It seems in fact that the method of logarithmic convexity cannot be applied directly to the Cauchy problem of linear anisotropic elastostatics.

BIBLIOGRAPHY

1. John, F., Continuous dependence on data for solutions of partial differential equations with a prescribed bound, Comm. Pure. Appl. Math. vol. 13 (1960) pp. 551-585.

2. Knops, R.J., Instability and the ill-posed Cauchy problem in elasticity, Mechanics of Solids (to appear).

3. Knops, R.J., Levine, H.A. and Payne, L.E., Nonexistence, instability and growth theorems for solutions of a class of abstract nonlinear equations with applications to nonlinear elastodynamics, Arch. Rat. Mech. Anal. vol. 55 (1974) pp. 52-72.

4. Knops, R.J. and Payne, L.E., Continuous data dependence for the equations of classical elastodynamics, Proc. Camb. Phil. Soc. vol. 84 (1978) pp. 177-190.

5. Knops, R.J. and Payne, L.E., Uniqueness and continuous dependence of the null solution in the Cauchy problem for a nonlinear elliptic system, Math. Research, G. Anger editor, vol. 1 (1979) pp. 151-160.

DEPARTMENT OF MATHEMATICS
CORNELL UNIVERSITY
ITHACA, NEW YORK 14853

Contemporary Mathematics
Volume 4
1981

A Nonlinear Eigenvalue Problem with an Exponential Nonlinearity

James Moseley

ABSTRACT. We consider the two dimensional nonlinear eigenvalue problem of Dirichlet type

$$\begin{aligned} -\Delta u &= \lambda(e^{u} + \Psi e^{-u}) & y \varepsilon \Omega \\ u &= \phi(y) & y \varepsilon \partial\Omega \end{aligned} \qquad (P)$$

where $y = (y_1, y_2)$, Δ is the Laplacian operator, Ψ is an analytic function of y, Ω is a simply connected region bounded by a smooth closed Jordan curve, the boundary data is continuous, and the parameter, λ, is positive. Problems of this type can arise in elasticity theory and combustion theory. This paper deals with obtaining a "large norm" (second) solution for (P) as $\lambda \to 0^+$. The procedure is as follows. First, an asymptotic solution is developed for a conformal transplantation of the problem to the unit disk. Then a modified Newton's method is developed for an equivalent integral equation formulation. For λ sufficiently small the method is shown to converge to a large norm solution provided the large norm asymptotic solution is used as a first approximation.

1. INTRODUCTION. In classical elasticity one encounters the case when the body forces depend on the hydrostatic pressure or on the direct stress within the material. In the two dimensional case, the compatibility equations assume the form:

$$\frac{\partial^2 \sigma}{\partial x^2} + \frac{\partial^2 \sigma}{\partial y^2} = -\frac{1}{2(1-\nu)} (k_1 x(\sigma) + k_2 y(\sigma))$$

where $\sigma = 1/2\ (\sigma_x + \sigma_y)$; that is σ is the average direct stress. Thus the compatibility conditions take the form

$$\Delta\sigma = \lambda\ f(\sigma)$$

where λ is a constant.

In wave propagation in an elastic medium one encounters the linear equation:

$$K \text{ grad } e + G\Delta u - \rho \frac{\partial^2 u}{\partial t^2} = 0.$$

1980 Mathematics Subject Classification 35J , 73H.

If grad e ≈ 0 this results in the eigenvalue problem

$$G\,\Delta u = \lambda u.$$

Investigating surface waves, Lord Rayleigh remarked about these waves, "diverging in two dimensions only, they must acquire at a great distance from the source a continuously increasing preponderence." It has been suggested that the inclusion of an exponential term in the corresponding eigenvalue problem would account for rapidly diminishing amplitude with the depth of the medium [18].

Nonlinear eigenvalue problems with exponential nonlinearities also arise in the theory of diffusion and reaction of gases [1], [9], [12], [7] (e.g. the Frank-Kamenetskiǐ model [8]); the theory of stellar structures [4], and the study of Riemann surfaces with bounded Gaussian curvature [11].

We begin by considering the problem

$$\begin{aligned} -\Delta u &= \lambda e^{u} + \nu e^{-u} && y \in \Omega \\ u(y) &= \phi(y) && y \in \partial\Omega \end{aligned} \qquad \text{(p)}$$

where $y = (y_1, y_2)$, Δ is the two dimensional Laplacian operator, Ω is a simply connected region bounded by a Jordan curve, the boundary data $\phi(y)$ is continuous on $\partial\Omega$, and ν and λ are real valued parameters.

There are two noteworthy cases depending on the signs of the parameters λ and ν.

Case A: $\lambda \leq 0$ $\nu \geq 0$

Case B: $\lambda > 0$

By letting $u(y) = v(y)$, $\lambda = -\nu_1$, and $\nu = -\lambda_1$ we obtain the equivalent problem

$$\begin{aligned} -\Delta v &= \lambda_1 e^{v} + \nu_1 e^{-v} && \Omega \\ v &= -\phi && \partial\Omega \end{aligned} \qquad \text{(P')}$$

which is also of type (P). The case $\lambda \leq 0$, $\nu < 0$ for (P) is equivalent to the case $\lambda_1 > 0$, $\nu_1 \geq 0$ for (P') and is therefore covered by Case B.

It is well known that there exists a unique solution for Case A ([5], p. 323, 372). In [17] a Newton type method is given for Case A which converges quadratically for any choice of the initial data.

Case B is a special case of

$$\begin{aligned} -\Delta u &= \lambda(e^{u} + \psi e^{-u}) && y \in \Omega \\ u &= \phi && y \in \partial\Omega \end{aligned} \qquad \text{(P1)}$$

where ψ is a real analytic function of y on Ω and continuous on $\overline{\Omega}$ and the assumption $\lambda > 0$ is part of the hypothesis.

For λ sufficiently small, a procedure for obtaining a sequence of functions which converges to the "low norm" solution (the solution which approaches the solution for the linear Dirichet problem for $\lambda = 0$) is well known ([5], p. 373). The purpose of the paper is to summarize the results of [20], [15], and [16] which develop a technique for obtaining a sequence of functions which converges to a "large norm" solution for λ sufficiently small.

2. SUMMARY OF RESULTS FROM THE LITERATURE. The outstanding feature of (P1) is that $-\Delta u = \lambda(e^{u} + \psi e^{-u})$ is analytic so that all solutions are analytic in Ω [13]. If in addition, the boundary $\partial\Omega$ and the boundary data ϕ are analytic, all solutions to the boundary value problem are analytic in $\bar{\Omega}$ [13]; however, this is too restrictive for our purposes.

To simplify our analysis, we let $u_1 = u - u_0$ where u_0 is the solution of the associated linear problem ($\lambda = 0$) and rewrite (P1) as

$$\begin{aligned} -\Delta u_1 &= \lambda(e^{u_0+u_1} + \psi e^{-(u_0+u_1)}) & y \in \Omega \\ u_1 &= 0 & y \in \Omega. \end{aligned} \qquad \text{(P2)}$$

If $\psi(y) \geq 0$ for $y \in \bar{\Omega}$ then all solutions of (P2) are super harmonic and hence positive.

If $0 \leq \psi(y) < e^{2u_0}$ we can apply the technique of Keller and Cohen [12] for points λ in the "spectrum" (the set of all $\lambda > 0$ such that (P2) has a solution) to obtain the "minimal" (low norm) solution. Keller and Cohen's results also show that for this case the spectrum is a finite interval with least upper bound λ^* satisfying $\lambda^* \leq \mu_1$ where μ_1 is the least eigenvalue of the linear problem

$$\begin{aligned} -\Delta v &= \mu e^{u_0} v & y \in \Omega \\ v &= 0 & y \in \partial\Omega. \end{aligned} \qquad \text{(L1)}$$

Bandle's results [2] show, in the case $0 \leq \psi(y) < e^{2\check{u}_0}$ where $\check{u}_0 = \min\limits_{y \in \Omega}(u_0(y))$, that for some $\varepsilon > 0$

$$\lambda^* \geq K_0 \frac{4}{A} + \varepsilon$$

where $K_0 = \max\limits_{m \geq 0}(m(e^{m+\hat{u}_0} + \hat{\psi} e^{-m-\check{u}_0})^{-1})$, $\hat{u}_0 = \max\limits_{y \in \Omega}(u_0(y))$, $\hat{\psi} = \max\limits_{y \in \Omega}(\psi(y))$,

and A is the area of Ω.

For $\phi \equiv 0$ and $\psi \equiv 0$, she has also shown [3] that

$$\frac{2\pi}{A} \leq \lambda^* \leq \frac{2}{R_0^2}$$

where R_0 is the maximal conformal radius of Ω. For this case she gives the extimates

$$1 - \sqrt{1 - \frac{\lambda}{2} R_y^2} \leq 2e^{u(y)/2} \leq 1 + \sqrt{1 - \frac{\lambda}{2} R_y^2}$$

where R_y is the conformal radius of y with respect to Ω.

Crandall and Rabinowitz's results [6] show that if $0 \leq \psi(y) < e^{2u_0}$, then λ^* is in the spectrum and there are two solutions for every $\lambda \in (0,\lambda^*)$.

For the case $\phi \equiv 0$, $\psi \equiv 0$, and $\partial\Omega$ a circle, the symmetric solutions of (P2) are given by:

$$e^{u_0} = \frac{b_0}{(1 + b_0\lambda r^2/8)^2} \quad \text{with } b_0 = \frac{32}{\lambda^2 R^4}\left[1 - \frac{\lambda R^2}{4} - \sqrt{1 - \frac{\lambda R^2}{2}}\right] \tag{2.1}$$

$$e^{u_1} = \frac{b_1}{(1 + b_1\lambda r^2/8)^2} \quad \text{with } b_1 = \frac{32}{\lambda^2 R^4}\left[1 - \frac{\lambda R^2}{4} + \sqrt{1 - \frac{\lambda R^2}{2}}\right] \tag{2.2}$$

where R is the radius of the circle and r is the radial variable [2]. In this case $\lambda^* = 2/R^2$ and it is easy to see that u_0 is the minimal solution, u_1 "blows up" at the origin as $\lambda \to 0$ and the two solutions coincide at λ^*. By a technique described in [10] it can be shown that these are the only solutions.

3. SUMMARY OF PROCEDURE. In order to obtain a "large norm" solution for (P1) we use the following procedure:

A. Obtain an equivalent problem (P2) with zero boundary data.

B. Obtain a conformal transplantation (P3) of (P2) to the unit disk.

C. Develop a "large norm" asymptotic ($\lambda \to 0^+$) solution for (P3) to order three.

D. Convert to an equivalent integral equation formulation.

E. Develop a modified Newton's method for the integral equation which converges to a "large norm" solution if the asymptotic solution given in C is used as a first approximation.

Step A was carried out in Section 2. Steps B through E will be carried out in Sections 4 through 7. Section 8 gives a special case for which a second "large norm" solution can be obtained. In Section 9 we summarize our results as Theorems 4 and 5. Some applications of these theorems are then given as corollaries.

4. CONFORMAL TRANSPLANTATION OF (P2). As no difficulty will arise, we will consider a region $\Omega \subseteq \mathbb{R}^2$ to also be a subset of $\mathbb{C}$ throughout this paper, using $y = (y_1, y_2) \in \Omega \subseteq \mathbb{R}^2$ and $w = y_1 + iy_2 \in \Omega \subseteq \mathbb{C}$. When the region is the unit disc U, we will use $x = (x_1, x_2) \in U \subseteq \mathbb{R}^2$ and $z = x_1 + ix_2 \in \mathbb{C}$. For any region $\Omega \subseteq \mathbb{C}$ with smooth boundary, we let $H(\Omega)$ be the holomorphic functions on Ω, and $A(\Omega)$ be the functions in $H(\Omega)$ which are continuous on $\bar{\Omega}$.

If f_Ω is any conformal mapping of the unit disk onto Ω with inverse mapping g_Ω, then the conformal transplantation of (P2) is given by:

$$\begin{aligned} -\Delta u_1 &= \lambda |f'_\Omega|^2 (e^{u_1 + u_0} + \psi e^{-u_1 - u_0}) & |x| < 1 \\ u_1 &= 0 & |x| = 1 \end{aligned} \qquad \text{(P3)}$$

where

$$u_1(x) = u_1(g_\Omega(w))$$

$$u_1(y) = u_1(f_\Omega(z)).$$

(No confusion will arise from this abuse of notation.)

To obtain asymptotic solutions for (P3) we let $h(w) = u_0(y) + iv_0(y)$ where u_0 is the solution of the associated linear problem $(\lambda = 0)$ and v_0 is its conjugate harmonic function. Next we let $f'_\phi(z) = \exp\{(1/2)h(f_\Omega(z))\}$ (the choice of notation is to emphasize that f'_ϕ plays a role similar to that of f'_Ω and we make no use of antiderivatives of f'_ϕ). The function of f'_ϕ characterizes the boundary data whereas f'_Ω characterizes Ω (except for translations). Hence (P3) is equivalent to

$$\begin{aligned} -\Delta u_1 &= \lambda |f'_\phi f'_\Omega|^2 (e^{u_1} + \psi |f'_\phi|^{-4} e^{-u_1}) & |x| < 1 \\ u_1 &= 0 & |x| = 1 \end{aligned} \qquad \text{(P4)}$$

where

$$u_1(x) = u(f_\Omega(z)) - \ln|f'_\phi(z)|^2$$

and

$$f'(z) \neq 0 \quad \text{for} \quad |z| < 1.$$

We refer to f, where $f' = f'_\phi f'_\Omega$, as an associated map for (P1) and note that it is not unique but depends on the choice of the conformal map f_Ω. However, the form of (P4) is independent of this choice.

5. NONUNIQUENESS OF THE ASSOCIATED MAP. In this section we consider the nonuniqueness of $f' = f'_\Omega f'_\phi$. We begin by defining for $|\tau| < 1$ the functions

$$\rho_\tau(z) = \frac{z - \tau}{1 - \bar{\tau} z} \qquad \rho_{-\tau}(z) = \frac{z + \tau}{1 + \bar{\tau} z}$$

which are inverse conformal maps of $\bar{U}$ onto $\bar{U}$.

It can be shown that for a given characterization, f_0 ($f_\Omega = f_{\Omega,0}$, $f_\phi = f_{\phi,0}$, and $f'_0 = f'_{\Omega,0} f'_{\phi,0}$) of a region Ω with boundary data ϕ, we can construct a family of characterizations via

$$f_{\Omega,\tau} = f_{\Omega,0} \circ \rho_{-\tau}$$

$$f'_{\phi,\tau} = f'_{\phi,0} \circ \rho_{-\tau}$$

$$f_\tau = f_0 \circ \rho_{-\tau}$$

with $f' = f'_{\Omega,\tau} f'_{\phi,\tau}$. Any of these characterizations is acceptable when formulating (P4). Also, since solutions of (P4) are independent of $f(0)$ and the argument of $f'(0)$, we may always reformulate to obtain $f(0) = 0$ and $f'(0) > 0$.

6. AN ASYMPTOTIC SOLUTION. To obtain an asymptotic solution for (P4), we first consider the case $\psi \equiv 0$. The general integral for $-\Delta u = \lambda |f'|^2 e^u$ for $|x| < 1$ can be written in the form

$$e^{-u/2} = |v(z)|^2 + \frac{\lambda}{8} \left| v(z) \int \frac{f'(\hat{z})d\hat{z}}{(v(\hat{z}))^2} \right|^2 \tag{6.1}$$

provided $v(z)$ has at most simple zeros in U and if $v(z_0) = 0$, then

$$\frac{v''(z_0)}{v'(z_0)} = \frac{f''(z_0)}{f'(z_0)} \tag{6.2}$$

Thus for any v, (6.1) gives a solution of $-\Delta u = \lambda |f'|^2 e^u$ where the requirement (6.2) renders the singularities of (6.1) to be removable. We may obtain an asymptotic solution (for $\psi \equiv 0$) by applying the boundary condition $u \equiv 0$ for $|x| = 1$ asymptotically; that is, requiring $u = O(\lambda^n)$ for $|x| = 1$. To obtain behavior similar to that exhibited by (2.2) we take

$$v(z;\lambda) = \frac{v_0(z)}{G(z;\lambda)} \tag{6.3}$$

where

$$G(z;\lambda) = 1 + \lambda\, G_1(z) + \dots \lambda^{n-1} G_{n-1}(z). \tag{6.4}$$

Thus to first order, we require v to be an inner function which we take to have a single zero;

$$v_0(z) = \frac{z - \delta}{1 - \bar{\delta} z} \tag{6.5}$$

Hence to first order (6.2) requires

$$\frac{f''(\delta)}{f'(\delta)} = \frac{2\bar{\delta}}{1 - |\delta|^2} \tag{6.6}$$

The existence of $\delta \in U$ satisifying (6.6) is assured if f' is bounded on U. However, to insure that the asymptotic solution (6.1) is in $C(U)$, the space of continuous functions on $\bar{U}$, we will require that $f' \in A(U)$ and $f'(z) \neq 0$ for $z \in \bar{U}$.

We now take advantage of the nonuniqueness of f. If $(f,\delta) = (f_0, \delta_0)$ satisfies (6.6) then (f_τ, δ_τ) where $f_\tau = f_0 \circ \rho_\tau$ and $\delta_\tau = \rho_{-\tau}(\delta_0)$ does also. Thus, since $f_\tau(\delta_\tau) = f_0(\delta_0)$ for all $\tau \in U$, $(f_{\delta_0}, 0)$ satisfies (6.6); that is $f''_{\delta_0}(0) = 0$.

To simplify further computation of the asymptotic solution (as well as the modified Newton's method), if (f,δ) satisfies (6.6) we normalize f by

$$f_n(z) = e^{i\beta}[f(\rho_{-\delta}(z)) - f(\rho_{-\delta}(0))] \tag{6.7}$$

where β is the argument of $\frac{d}{dz}(f(\psi_{-\delta}(z)))$ evaluated at zero. Hence $f_n(0)$ and $f'_n(0) > 0$ as well as $f''_n(0) = 0$.

Thus it will suffice to consider

$$\begin{array}{lll} \text{PDE} & -\Delta u = \lambda|f'(z)|^2(e^u + \sigma(x)e^{-u}) & |x| < 1 \\ \text{BC} & u = 0 & |x| = 1 \end{array} \tag{P5}$$

where $f \in A(u)$, $f'(z) \neq 0$ for $z \in \bar{U}$, $f(0) = 0$, $f'(0) > 0$, $f''(0) = 0$, and $\sigma(x)$ is a real analytic function of $x = (x_1, x_2)$ in U and continuous on $\bar{U}$.

To obtain an asymptotic solution of (P5) we choose u to be of the form

$$u(z;\lambda) = u_e(z;\lambda) + H(z;\lambda) \tag{6.8}$$

where

$$H(z;\lambda) = \lambda H_1(z) + \lambda^2 H_2(z) + \dots + \lambda^{n-1} H_{n-1}(z)$$

where $u_e(z;\lambda)$ is given by

$$e^{-u_e/2} = \frac{|z|^2 + \frac{\lambda}{8}\left|z \int^z G(\hat{z};\lambda) \frac{f(\hat{z})}{\hat{z}^2} d\hat{z}\right|^2}{|G(z;\lambda)|^2}, \tag{6.9}$$

with G being given by (6.3). In the notation of [15] this may be written as

$$e^{-u_e/2} = \frac{|z|^2 + (\frac{\lambda}{8})N}{K}.$$

Substituting (6.8) into the PDE given in (P5) yields

$$-\Delta H = \lambda|f'|^2(e^{u_e}(e^H - 1) + \sigma e^{-u_e}e^{-H}).$$

This can be shown to be equivalent to $E(z:\lambda) = 0$ where

$$E(z:\lambda) =$$

$$(\Delta H)(|z|^2 + \tfrac{\lambda}{8}N)^2K^2 + \lambda|f'|^2(K^4(e^H - 1) + \sigma(|z|^2 + \tfrac{\lambda}{8}N)^4e^{-H}). \quad (6.10)$$

If $\|E\| = \max_{|z|\leq 1} |E(z:\lambda)| = O(\lambda^n)$ and $\max_{|z|=1} |u(x:\lambda)| = O(\lambda^n)$ we will say that $u(x:\lambda)$ is an asymptotic solution of (P5) of order n. We will see that a solution of order 3 is sufficient. Hence we need only compute H_1, G_1 and H_2, G_2.

Equating terms of order up to three yields that H_i, $i = 1,2$ must satisfy

$$-\Delta H_i = Q_i(z) \qquad i = 1,2 \qquad (6.11)$$

where $Q_1 = \sigma|f'|^2 |z|^4$ and Q_2 depends on H_1 and G_1 as well as σ and f' and is given in [15]. The requirement that σ be a real analytic function of x is sufficient to insure that H_1 and H_2 can be chosen to be in $C(U)$. Applying the boundary condition up to order three requires

$$2\mathrm{Re}G_i = \frac{1}{8}(P_{i-1}(z) + R_{i-1}(z)) \qquad i = 1,2 \qquad (6.12)$$

where P_0 and R_0 depend on f', σ, and H_1 and P_1 and R_1 depend on f', σ, H_1, G_1, and H_2 as given in [15]. Hence we may choose successively H_1, G_1, H_2 and G_2. However, the requirement (6.2) implies $G_i'(0) = 0$. This can be satisfied by appropriate choice of the integration constant (which is expanded asymptotically) provided

$$|f'''(0)| \neq 2|f'(0)|. \qquad (6.13)$$

Summarizing, we have

THEOREM 1. If (6.13) is satisfied, a large norm asymptotic solution of (P5) can be obtained in the form (6.8) to order three by choosing successively H_1, G_1, H_2, and G_2 as indicated above. That is:

$$\max_{|z|\leq 1} |E(z;\lambda)| = O(\lambda^3)$$

$$\max_{|z|=1} |u(z:\lambda)| = O(\lambda^3)$$

where $E(z;\lambda)$ is given by (6.10) and

$$\max_{|z|\le 1} |u(z:\lambda)| = O\left(\ln \frac{1}{\lambda}\right)$$

7. A MODIFIED NEWTON'S METHOD. In this section we will show how the large norm asymptotic solution for (P5) given in the previous section can be used to generate an exact solution via a modified Newton's iteration scheme for an equivalent integral equation formulation.

We convert (P5) to the equivalent integral equation ($u \in C(U)$) with $\|u\| = \max_{|x|\le 1} |u(x)|$)

$$u = \mathbb{K}u$$

where

$$\mathbb{K} = \mathbb{K}_1 + \mathbb{K}_2$$

$$(\mathbb{K}_1 u)(x_0) = \int_{|x|\le 1} g(x_0,x)\lambda e^{u(x)} |f'(x)|^2 dx$$

$$(\mathbb{K}_2 u)(x_0) = \int_{|x|\le 1} g(x_0,x)\lambda\sigma(x) e^{-u(x)} |f'(x)|^2 dx.$$

We use a modified Newton's method of the form

$$u_{n+1} = S(u_n) \tag{7.1}$$

where

$$Su = (\mathbb{I} - \mathbb{K}'_{u_0})^{-1}(\mathbb{K}u - \mathbb{K}'_{u_0}u)$$

and $\mathbb{K}'_{u_0}$ is the Fréchet derivative of $\mathbb{K}$ evaluated at u_0. That is,

$$\mathbb{K}'_{u_0} = \mathbb{K}'_{1u_0} + \mathbb{K}'_{2u_0} \tag{7.2}$$

where

$$(\mathbb{K}'_{1u_0} h)(x_0) = \int_{|x|\le 1} g(x_0,x)\lambda e^{u_0(x)} h(x) |f'(z)|^2 dx$$

$$(\mathbb{K}'_{2u_0} h)(x_0) = \int_{|x|\le 1} g(x_0,x)\lambda\sigma(x) e^{-u_0(x)} h(x) |f'(z)|^2 dx.$$

Hence

$$S'_u h = (\mathbb{I} - \mathbb{K}'_{u_0})^{-1} (\mathbb{K}'_u h - \mathbb{K}'_{u_0} h)$$

and since

$$\mathbb{K}'_u h - \mathbb{K}'_{u_0} h = \mathbb{K}'_{1u_0}[(e^{u-u_0} - 1)h] + \mathbb{K}'_{2u_0}[(e^{-(u-u_0)} - 1)h]$$

we have that

$$\| S'_u \| \leq \| (\mathbb{I} - \mathbb{K}'_{u_0})^{-1} \mathbb{K}'_{1u_0} \| \, \| e^{u-u_0} - 1 \|$$

$$+ \| (\mathbb{I} - \mathbb{K}'_{u_0})^{-1} \mathbb{K}'_{2u_0} \| \, \| e^{-(u-u_0)} - 1 \|$$

Hence it can be shown that

$$\| S'_u \| \leq \Gamma(e^t - 1) \quad \text{if} \quad \| u - u_0 \| \leq t$$

where Γ satisfies

$$\| (\mathbb{I} - \mathbb{K}'_{u_0})^{-1} \mathbb{K}'_{1u_0} \| + \| (\mathbb{I} - \mathbb{K}'_{u_0})^{-1} \mathbb{K}'_{2u_0} \| \leq \Gamma. \tag{7.3}$$

Now since

$$S(u) = u - (\mathbb{I} - \mathbb{K}'_{u_0})^{-1} (u - \mathbb{K}u)$$

$$\| (\mathbb{I} - \mathbb{K}'_{u_0})^{-1} \| \leq 1 + \| (\mathbb{I} - \mathbb{K}'_{u_0})^{-1} \mathbb{K}'_{u_0} \|$$

we also have

$$\| S(u_0) - u_0 \| \leq (1 + \Gamma) \| u_0 - \mathbb{K}_{u_0} \|.$$

Hence

$$\phi(t) = \Gamma(e^t - t - 1) + \phi(0)$$

majorizes S [19, p.260] provided that $\phi(0)$ satisfies

$$(1 + \Gamma) \| u_0 - \mathbb{K}u_0 \| \leq \phi(0).$$

It is easy to show that $\phi(t) = t$ has a unique positive solution $t^* \leq \ln[1 + (1/\Gamma)]$ if

$$\| u_0 - \mathbb{K}u_0 \| \leq \ln[1 + (1/\Gamma)] - \frac{1}{1 + \Gamma}.$$

Hence applying the result of Kantorovich [19, p.260] we have the following:

THEOREM 2. If $\| u_0 - Ku_0 \| \leq \ln[1 + (1/\Gamma)] - 1/(1 + \Gamma)$ where Γ satisfies (7.3), then the modified Newton's method (7.1) will converge to a solution u^* of (P5) such that

$$\| u_0 - u^* \| \leq \ln[1 + (1/\Gamma)].$$

It remains to estimate $\| u_0 - \mathbb{K}u \|$ and Γ. In [15] we obtain the following estimates

$$\| u_0 - \mathbb{K}u_0 \| = O(\lambda^{5/2} \ln^{1/\lambda})$$

$$\| \mathbb{K}'_{1u_0} \| = O(\ln\tfrac{1}{\lambda}) \qquad \| K'_{2u_0} \| = O(\lambda)$$

$$\| (\mathbb{I} - \mathbb{K}'_{u_0})^{-1} \| = O(\tfrac{1}{\lambda})$$

Hence $\Gamma = O(\frac{\ln \frac{1}{\lambda}}{\lambda})$.

Recalling that for $\Gamma > 1$ we have

$$\ln\left(1 + \frac{1}{\Gamma}\right) - \frac{1}{1 + \Gamma} = \frac{1}{2}\left(\frac{1}{\lambda}\right)^2 - \frac{2}{3}\left(\frac{1}{\lambda}\right)^3 + \dots$$

and that

$$\lambda^{5/2} \ln \frac{1}{\lambda} = o\left(\lambda^2 \ln\left(\frac{1}{\lambda}\right)^{-2}\right) .$$

we apply Theorem 2 to obtain:

THEOREM 3. If the large norm asymptotic solution (6.8) to order three is used as a first approximation in the modified Newton's method (7.1), then u_n will converge to a unique large norm solution u^* such that

$$\| u_0 - u^* \| = O(\lambda).$$

8. A THIRD SOLUTION OF (P1) IF ψ IS A NEGATIVE CONSTANT. Suppose that ψ is a negative constant and that λ is sufficiently small so that we may obtain a large norm "positive" solution to both (P1) and

$$\begin{aligned} -\Delta u &= \lambda(-\psi)(e^u + (1/\psi)e^{-u}) \\ u &= -\phi \end{aligned} \qquad \text{(P6)}$$

by the technique of the previous sections. We call such solutions "positive" since at the point $f_\Omega(\delta)$ the solution increases positively without bound as $\lambda \to 0$. If $\hat{u}$ is the solution of (P6) obtained, then $-\hat{u}$ is a large norm "negative" solution of (P1).

9. SUMMARY OF RESULTS. We summarize our results:

THEOREM 4. If $|f'''(0)| \neq 2|f'(0)|$, then an asymptotic solution of (P5) of order three can be obtained which will generate, for λ sufficiently small, an exact large norm (second) solution via a modified Newton's iteration scheme for an equivalent integral equation. The asymptotic solution is given by

$$u_A = u_e(z;\lambda) + H(z;\lambda)$$

where

$$e^{\frac{-u_e(z;\lambda)}{2}} = \frac{|z|^2 + \frac{\lambda}{8}\left|z\int^z [G(\hat{z};\lambda)]^2 \frac{f'(\hat{z})}{z^2}\, d\hat{z}\right|^2}{|G(z;\lambda)|^2}$$

$$G(z;\lambda) = 1 + \lambda G_1(z) + \lambda^2(G_2(z))$$

$$H(z;\lambda) = \lambda H_1(z) + \lambda^2 H_2(z);$$

the functions G_i and H_i being as in Section 6.

THEOREM 5. Let the normalized associated map f_n of (P1) satisfy $f_n \in A(u)$, $f_n(z) \neq 0$ for $z \in \bar{u}$, $f_n(0) = 0$, $f_n'(0) > 0$, $|f_n'''(0)| \neq 2\, f_n(0)$, and $f_n' = f_\Omega' f_\phi'$. Furthermore, let g_Ω be the inverse of f_Ω, u_A be the asymptotic solution of (P5) for $\sigma = \psi|f_\phi'|^{-\psi}$, and u_F be the exact solution generated by the Newton's method in Theorem 4. Then a large norm asymptotic solution of (P1) is given by $u_A(g_\Omega(w)) + \ln|f_\phi'(g_\Omega(w))|$ and an exact solution is given by $u_F(g_\Omega(w)) + \ln|f_\phi'(g_\Omega(w))|$.

We call $\partial\Omega$ Dini-smooth if (in addition to being smooth) the tangent angle $\beta(s)$ as a function of arclength s satisfies $|\beta(s_2) - \beta(s_1)| < \omega(s_2 - s_1)$ for $s_1 < s_2$ where $\omega(x)$ is an increasing function for which $\int_0^1 \omega(x)/x\, dx < \infty$. If $\partial\Omega$ is Dini-smooth then $f_\Omega'(z) \in A(U)$, $f_\Omega'(z) \neq 0$ for $z \in \bar{U}$, and $g_\Omega'(w) \in A(\Omega)$ where g_Ω is the inverse of f_Ω' [16, p.298]. Note that if $\beta(s)$ satisfies a Lipschitz condition of order α, $0 < \alpha \leq 1$, it is Dini-smooth. Clearly $f_\phi'(z) \neq 0$ on $\bar{U}$. It can be shown that if β satisifies a Lipschitz condition of order α, $0 < \alpha \leq 1$, on $\partial\Omega$, then $f_\phi' \in A(U)$.

COROLLARY 1. If $\phi \equiv 0$, Ω is convex, and $\partial\Omega$ is Dini-smooth, then the conclusion of Theorem 5 is valid.

In the following corollaries we always assume that $\partial\Omega$ is Dini-smooth and that $h = u_0 + iv_0$ is in $A(\Omega)$ where $u_0(w) = \phi(w)$ for $w \in \partial\Omega$. Hence we we always have $f' = f_\Omega' f_\phi' \in A(U)$ and $f'(z) \neq 0$ for $z \in \bar{U}$.

COROLLARY 2. In addition to the above smoothness requirements, assume that f'_Ω and f'_ϕ have been normalized separately (i.e., $f''_\Omega(0) = f''_\phi(0) = 0$) and that Ω is convex. If $|f''_\phi(0)/f'_\phi(0)| < 2/M^2$ where $M = \max_{z\varepsilon\bar{U}} |f'_\Omega(z)/f'_\Omega(0)|$, then the conclusion of Theorem 5 is valid.

COROLLARY 3. In addition to the above smoothness requirements, assume that Ω is convex, that $\partial\Omega$ is symmetric with respect to two perpendicular axes, and that f_Ω is the odd conformal map of U onto Ω with $f_\Omega(0) = 0$ and $f'_\Omega(0) > 0$. If ϕ is symmetric with respect to both axes, and

$$|h''(0)| < \frac{4}{M^2} \tag{9.1}$$

where $M = \max_{z\varepsilon\bar{U}} |f'_\Omega(z)|$, then the conclusion of Theorem 5 is valid.

COROLLARY 4. The requirement (9.1) in Corollary 3 may be replaced by $M_0M^2 < 4|f'_\Omega(0)|$ when $M_0 = \max_{w\varepsilon\bar{\Omega}} |h'(w)| < \infty$.

REFERENCES

1. Aris, R., The Mathematical Theory of Diffusion and Reaction in Permeable Catalysts, Clarendon Press, Oxford, 1975.

2. Bandle, C., Existence theorems, qualitative results and a priori bounds for a class of nonlinear Dirichlet problems, Arch. for Rat. Mech. and Anal., 58(1975), 219-238.

3. __________, Isoperimetric inequalities for a nonlinear eigenvalue problem, Proc. Amer. Math. Soc., 56(1976), 243-246.

4. Chandrasekar, S., An Introduction to Stellar Structures, Dover, New York, 1957.

5. Courant, R. and D. Hilbert, Methods of Mathematical Physics, Vol. II, Interscience, New York, 1962.

6. Crandall, M. G. and P. N. Rabinowitz, Some continuation and variational methods for positive solutions of nonlinear elliptic eigenvalue problems, Arch. for Rat. Mech. and Anal., 58(1975), 207-218.

7. Fradkin, L., Jr. and G. C. Wake, Nonlinear eigenvalue problems, Bull. Austral. Math. Soc., Vol. 12 (1975, 467-472.

8. Frank-Kamenetskii, D. A., Diffusion and Heat Transfer in Chemical Kinetics, Plenum Press, New York, 1969.

9. Gel'fand, I. M., Some problems in the theory of quasilinear equations, Amer. Math. Soc. Trans., 29(1963), 295-381.

10. Gidas, B., Wei-Ming Ni, and L. Nirenberg, Symmetry and related properties via the maximum principle, Com. Math. Phy., 68(1979), 209-243.

11. Kazdan, J. L. and F. W. Warner, Curvature functions for compact 2-manifolds, Ann. of Math., 99(1974), 14-47.

12. Keller, J. B. and D. S. Cohen, Some positone problems suggested by nonlinear hear generation, J. Math. Mech., 16(1967), 1361-1376.

13. Morrey, C. B., On the analyticity of the solutions of analytic nonlinear elliptic systems of partial differential equations, I and II, A. J. Math., 80(1958), 198-237.

14. Moseley, J. L., Asymptotic solutions for a Dirichlet problem with an exponential nonlinearity, submitted to SIAM J. Math. Anal. (1980).

15. ____________, A two dimensional Dirichlet problem with an exponential nonlinearity, to appear.

16. Pommerenke, C., Univalent Functions, Vandenhoeck & Ruprecht, Göttingen, 1975.

17. Schryer, N. L., Solution of monotone nonlinear elliptic boundary value problems, Numer. Math. 18(1972), 336-344.

18. Timoshenko, S. and J. N. Goodier, Theory of Elasticity, McGraw-Hill, New York, 1951.

19. Vaǐnberg, M. M., Variational Methods for the Study of Nonlinear Operators, Holden-Day, San Francisco, 1964.

20. Weston, V. H., On the asymptotic solution of a partial differential equation with an exponential nonlinearity, SIAM J. Math. Anal., 9(1978), 1030--1053.

DEPARTMENT OF MATHEMATICS
WEST VIRGINIA UNIVERSITY
MORGANTOWN, WV 26505

Contemporary Mathematics
Volume 4
1981

EIGENVALUE PROBLEMS FOR VARIATIONAL INEQUALITIES

Erich Miersemann

ABSTRACT. We obtain some results for the variational inequality $u \in V$: $a(u, v - u) \geq \lambda\, b\, (u, v - u)$, valid for all $v \in V$ where V is a closed convex subset of a Hilbert space. Applications of our theorems are given for problems in elastic stability, beam and plate theory. Specifically, buckling problems with constraints on buckling deflection, vibration problems and general stability problems lead to the specific formulation of the inequalities discussed here.

I. INTRODUCTION. In this paper we discuss some results for the eigenvalue problem

$$u \in V:\ a(u_1 v - u) \geqq \lambda b(u_1 v - u), \quad \text{for all} \quad v \in V, \tag{1}$$

where V is a closed convex subset of a real Hilbert space H. We assume that $0 \in V$. $a(u,v)$, $b(u,v)$ are real bilinear forms which are symmetric and defined on H. The eigenvalue parameter λ is real. For further assumptions see the following chapters.

Using the methods, which will be discussed in the following section, one may obtain results for the more general problem

$$u \in V:\ (f'(u)_1 v - u) \geqq \lambda(g'(u)_k v - u) \qquad \text{for all} \quad v \in V. \tag{2}$$

Here the f', g' are the first Fréchet derivatives of functionals f, g defined on H. For $u,v \in H$ let (u,v) be an inner product on H.

To simplify matters, some methods and proofs will be sketched for the problem (1) only.

Basis results for variational inequalities were given, for example, in Lions, Stampacchia [17]. But inequalities involving an eigenvalue parameter were investigated years later.

Without attempting to formulate any precise definitions, let us discuss a simple example. Consider a beam clamped at both ends and compressed by a force P. By $v(x)$ we denote the deflection of the beam away from the x-axis. The Euler critical load P_o is given by

Mathematics Subject Classification 49H05, 73H10.

$$P_o^{-1} = \max_{\substack{v\in L \\ v\neq 0}} \frac{\int_0^1 v'^2\, dx}{EJ \int_0^1 v''^2\, dx},$$

where $L = \{v/v(0) = v'(0) = v(1) = v'(1) = 0\}$, and EJ is the bending stiffness. The critical load P_o is the first eigenvalue of

$$EJu^{(4)} = -Pu'' \quad \text{in } (0,1)$$

$$u(0) = u'(0) = u(1) = u'(1) = 0.$$

Now let the deflections be restricted by obstacles. Define a convex set of functions

$$V = \{v|v \in L,\ \psi_1(x) \leqq v(x) \leqq \psi_2(x) \quad \text{on} \quad (0,1)$$

The quantities ψ_1, ψ_2 are given and satisfy $\psi_1(x) \leqq 0 \leqq \psi_2(x)$ on $(0,1)$; see Figure 1.

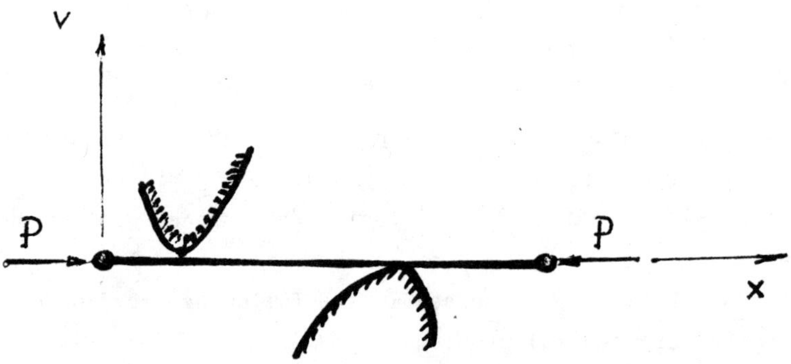

Figure 1

In this case it is also possible to define a critical load. The eigenfunctions are solutions of the variational inequality

$$u \in V: \quad EJ \int_0^1 u''(v - u)''dx \geqq P \int_0^1 u'(v - u)'dx,$$

valid for all $v \in V$.

In Chapter II we deal with the existence of solutions (λ,u), $u \neq 0$, of the variational inequality (1), and we give some bifurcation results for the first eigenvalue and an application to the plate theory.

The problem of higher eigenvalues and an example will be discussed in Chapter III.

In Chapter IV we consider eigenvalue equations in convex cones. We study conditions under which a solution (λ,u) of the inequality, where V is a convex and closed cone with vertex at zero, is in fact a solution to the associated equation. It can be shown that the problem

$$\Delta^2 u = -\lambda\Delta u \qquad \text{or} \qquad \Delta^2 u = \lambda u$$

for a smooth, convex, simply supported plate has a positive eigenfunction associated with the lowest simple eigenvalue.

In the last Chapter V we give a criterion that a solution to the inequality (1) defines a strong local minimum of an associated energy functional. Furthermore, we discuss the possibility of bifurcation fron nonzero solutions of variational inequalities. As an application of these, we consider a compressed beam where the admissible deflections v satisfy the inequality $|v| \leqq d$, $0 < d = \text{const}$. This problem was studied by Link [16] without the use of variational inequalities. The admissible deflections are defined as

$$V = \{v \mid v(0) = v(1) = 0,\ |v(x)| \leqq d \text{ in } (0,1)\}.$$

For $P > P_o$, here P_o denotes the critical load of Euler, the beam contacts the line $y = d$, as seen on Figure 2. There exists a critical value $P_{crit} > P_o$ where the beams deflection switches to another state.

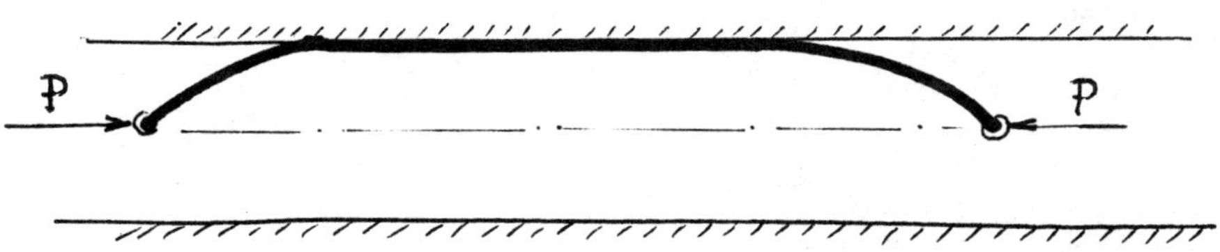

Figure 2

For a plate the same problem was treated in [25].

ACKNOWLEDGMENTS. I would like to express my gratitude to Professor Beckert for initiating my studies in variational inequalities and I would like to thank Professor Klötzler for telling me about the problem of Link [16].

II. FIRST EIGENVALUE, BIFURCATION. We start with a simple case. Let V be a cone with its vertex at zero, that is, a set such that $t \cdot u \in K$ for all $t > 0$ and for all $u \in K$. Furthermore, we assume that K is closed, convex and non-empty.
We look for solutions (λ, u), $\lambda \in \mathbb{R}$, $u \neq 0$, of the variational inequality (1). Under the assumptions

$$\begin{aligned} &a(u,u) \geqq 0 \quad \text{for all} \quad u \in H, \\ &\text{there exists} \quad c > 0 \quad \text{such that} \\ &a(v,v) \geqq c\,\|v\|^2 \quad \text{for all} \quad v \in V, \text{ and} \\ &b(u,v) \quad \text{is completely continuous on} \quad H \end{aligned}$$

We have :

THEOREM 1 [18]. *Suppose there exists a* $w \in K$ *such that* $b(w,w) > 0$. *Then the following maximum problem is solvable and* λ_0 *defined by*

$$\lambda_0^{-1} = \max_{\substack{v \in K \\ v \neq 0}} \frac{b(v,v)}{a(v,v)}$$

is the smallest positive eigenvalue of the variational inequality (1).

For the proof we use a method due to Beckert [1] and Krasnosel'skiĭ [12], which was generalized to inequalities by the author.

Let M_s be the set

$$M_s = \{v \in K \mid a(v,v) \leqq s\} \ , \quad \text{where} \quad 0 < s < \infty \ .$$

We seek vectors $u \in M_s$ such that

$$b(u,u) = \max_{v \in M_s} b(v,v).$$

From these assumptions it follows that this problem has a solution u, and that for every solution u we have $a(u,u) = s$. Let $v \in K$ be fixed and

$0 < \varepsilon < \varepsilon_o$, with ε_o sufficiently small. We calculate $k(\varepsilon)$ such that $a(w,w) = s$ for

$$w = (1 + k)[u + \varepsilon(v - u)].$$

We obtain

$$k(\varepsilon) = -\frac{a(u, v - u)}{a(u,u)} \varepsilon + o(\varepsilon).$$

Since $w \in M_s$ it follows that

$$b(w,w) \leqq b(u,u).$$

And from this inequality we deduce the inequality (1), where

$$\lambda = \frac{a(u,u)}{b(u,u)} .$$

The number λ is independent of the parameter s. This follows from the inequalities

$$\begin{aligned} \lambda^{-1}(s) &= \frac{b(u_s,u_s)}{a(u_s,u_s)} \\ &= \frac{b[(u_s/\sqrt{s}),(u_s/\sqrt{s})]}{a[(u_s/\sqrt{s}),(u_s/\sqrt{s})]} \\ &\leqq \lambda^{-1}(1) \\ &= \frac{b(\sqrt{s}u_1, \sqrt{s}u_1)}{a(\sqrt{s}u_1, \sqrt{s}u_1)} \\ &\leqq \lambda^{-1}(s) \quad , \end{aligned}$$

where u_r denotes a solution with respect to the parameter r of the above maximum problem.

REMARK. Suppose there exists a $w \in K$ such that $b(w,w) < 0$. Then λ_o defined by

$$-\lambda_o^{-1} = \max_{\substack{v\in K \\ v\neq 0}} - \frac{b(v,v)}{a(v,v)}$$

is the greatest negative eigenvalue of (1). The proof is the same as the one given above. In this case we pose the maximum problem $-b(v,v) = \max!$ in M_s.

Under certain assumptions it was proved in [18] that λ_o as given in Theorem 1 is the smallest positive point of bifurcation for an associated nonlinear problem of the type (2), where $V = K$. A different proof of this result, based on a penalty method, was given by Do [8].

Here λ is said to be a *point of bifurcation* if there exists a sequence of solutions (λ_n, u_n) of (2) where $u_n \neq 0$, $\lambda_n \to 0$ and $u_n \to 0$ as n tends to infinity.

Now, let V be a closed convex set of H with 0 V. Denote the $C(V)$ the tangential cone of V at zero (that means, the closure of the set

$$\{w = t \cdot v \quad \text{for all} \quad v \in V, \text{ for all} \quad t > 0\}.)$$

It is easy to see that $C(V)$ is a closed convex cone with vertex at zero.

Under the assumptions of Theorem 1, where $K = C(V)$, we have

THEOREM 2 [20]. *The number* λ_o *is the smallest positive point of bifurcation to the inequality* (1).

Under suitable assumptions λ_o is also a point of bifurcation for associated nonlinear problems of the type (2).

For any eigenvalue $\lambda > 0$ of (1) we have $\lambda \geqq \lambda_o$. Hence, from the inequality (1) it follows that for an eigenvector u corresponding to λ the inequality

$$\lambda^{-1} \leqq \frac{b(u,u)}{a(u,u)}$$

is true. This is checked by setting $v = 0$. Setting $u \in V \subset C(X)$ we get

$$\frac{b(u,u)}{a(u,u)} \leqq \lambda_o^{-1}.$$

EXAMPLE. The buckling problem for the elastic clamped plate with constraints on the variations leads to an unilateral eigenvalue problem. Assume that the set of variations $V \subset H_{2,2}(\Omega)$ is convex with $0 \in V$. The following inequality can be deduced from a stability criterion due to Trefftz [27] under von Karman's assumptions [11] (see [20]).

$$u \in V: \int_\Omega \Delta u \Delta(v - u)\, dx + \int_\Omega \sigma_{ij}(u) u_{x_i} (v - u)_{x_j}\, dx$$
$$\geqq \lambda \int_\Omega \sigma_{ij}(x) u_{x_i} ((v - u)_{x_j}\, dx \quad \text{for all} \quad v \in V. \tag{3}$$

We are using the summation convention. Here $\sigma_{ij} = \sigma_{ji}$ denote certain quadratic functionals, Ω a bounded and regular domain in $\mathbb{R}^2$ and $H_{2,2}(\Omega)$ is a well-known Sobolev space.

The bifurcation problem for (3) was investigated by Do [8] and by the author for the case when V is a closed convex cone with vertex at zero. The general case of a convex set one can find in [20].

The associated equation for (3) was studied by Berger [3] and Berger, Fife [4].

THEOREM 3 [23]. *We assume that for every* $w \in V$, $w \neq 0$, *one can find a vector* $v \in V$ *such that* $b(w, v - w) > 0$. *Then there exists for every* $0 < s < \infty$ *a solution* u *to the variational inequality* (1) *with* $a(u,u) = s$.

Let

$$M_s = \{v \in V \mid a(v,v) \leqq s\} \quad \text{where} \quad 0 < s < \infty \quad .$$

By using a lemma in [23] it follows that for each solution u of

$$b(u,u) = \max_{v \in M_s} b(v,v)$$

we have $a(u,u) = s$ and that there exists a $v \in V$ such that $a(u, v - u) > 0$ and $b(u, v - u) > 0$.

Let $v,z \in V$ be fixed with $a(u, z - u) \neq 0$ and let $0 < \varepsilon < \varepsilon_o$, with ε_o sufficiently small. We calculate $k(\varepsilon)$ such that $a(w,w) = s$ for

$$w = (1 - k)[u + \varepsilon(v - u)] + kz.$$

We obtain

$$k(\varepsilon) = - \frac{a(u,v - u)}{a(u,z - u)} \varepsilon + o(\varepsilon).$$

Let us set

$$C_u^+ = \{v \in V \mid a(u, v - u) > 0\} \quad \text{and} \quad C_u^- = \{v \in V \mid a(u, v - u) < 0\}.$$

Then for every solution u of the above maximum problem we have

$$C_u^+ \neq \emptyset \qquad \text{and} \qquad C_u^- \neq \emptyset.$$

If $z \in C_u^+$ and $v \in C_u^-$, we conclude that $0 < k(\varepsilon) < 1$, provided $\varepsilon_o > 0$ is sufficiently small. Hence, we have $w \in V$. Since $b(w,w) \leqq b(u,u)$, we deduce the inequality

$$\frac{b(u, z - u)}{a(u, z - u)} \leqq \frac{b(u, v - u)}{a(u, v - u)} \quad .$$

We denote

$$\alpha = \sup_{z\in C_u^+} \frac{b(u, z - u)}{a(u, z - u)} , \qquad \beta = \inf_{v\in C_u^-} \frac{b(u, v - u)}{a(u, v - u)}$$

Then u is a solution of the variational inequality (1) for all λ with $\lambda^{-1} \in [\alpha,\beta]$ and for all $v \in C_u^+ \cup C_u^-$.

In the case $a(u, v - u) = 0$ we put $v_n = (1 - (1/n))v$ into the inequality (1) and let $n \to \infty$. Since $v_n \in C_u^-$, the above inequality follows for v.

The corresponding nonlinear problem was treated in [23].

In order to illustrate Theorem 3 we consider a linear problem corresponding to (3):

$$u \in V: \int_\Omega \Delta u\Delta(v - u)\, dx \geqq \lambda \int_\Omega a_{ij}(x)u_{x_i}(v - u)_{x_j}\, dx \qquad \text{for all } v \in V,$$

where

$$V = \{v \in \overset{\circ}{H}_{2,2}(\Omega) \mid \psi_1(x) \leqq v(x) \leqq \psi_2(x) \text{ in } \Omega\} ,$$

with $\psi_1(x) \leqq 0 \leqq \psi_2(x)$ in Ω and $\psi_1, \psi_2 \in H_{2,2}(\Omega)$. We assume $a_{ij} = a_{ji}$ and $a_{ij} \in C^1(\bar{\Omega})$. We denote by

$$L = - \frac{\partial}{\partial x_j}\left(a_{ij}\frac{\partial}{\partial x_i}\right) .$$

Then the hypothesis of Theorem 3 is fulfilled if

i) $V \neq 0$.

ii) $L\psi_1 \geqq 0$, $L\psi_2 \leqq 0$ almost everywhere on Ω (that is, ψ_1 is a supersolution and ψ_2 a subsolution with respect to L).

iii) There exists $c > 0$ such that

$$a_{ij}(x)\xi_i\xi_j \geqq c|\xi|^2$$

for all $\xi \in \mathbb{R}^2$ and for all $x \in \Omega$.

If it is not true, we conclude from

$$\int_\Omega a_{ij}(x)u_{x_i}(v - u)_{x_j}\, dx \leqq 0, \quad \text{for all } v \in V,$$

that

$$Lu = \begin{cases} 0 & \text{if } \psi_1(x) < u(x) < \psi_2(x) \\ L\psi_1 & \text{if } u(x) = \psi_1(x) \\ L\psi_2 & \text{if } u(x) = \psi_2(x) . \end{cases}$$

Therefore we obtain

$$\int_{\Omega} a_{ij}(x)u_{x_i}u_{x_j}\,dx = \int_{u=\psi_1} L\psi_1\psi_1\,dx + \int_{u=\psi_2} L\psi_2\psi_2\,dx \leqq 0 .$$

which is impossible because $u \in V$ and $u \neq 0$. The inequality $u \neq 0$ follows from the assumptions i) and ii).

Eigenvalue problems for variational inequalities with nondifferentiable functionals were studied by Do in [9].

III. HIGHER EIGENVALUES. Let $V \subset H$ be a closed convex cone with the vertex at zero. We consider the equation

$$u \in H: \quad a(u,v) = \mu b(u,v) \qquad \text{for all} \quad v \in H \tag{4}$$

and for a closed linear subspace $H_1 \subset V$

$$u \in H_1: \quad a(u,v) = kb(u,v) \quad \text{for all} \quad v \in H_1. \tag{5}$$

We assume that for (5) there exist $n + 1$ positive eigenvalues

$$0 < K_1 \leqq K_2 \leqq \cdots \leqq K_{n+1} .$$

Since $H_1 \subset H$ it follows that there exist $n + 1$ positive eigenvalues μ_i of (4) with $0 < \mu_i \leqq K_i$.

THEOREM 4. *Suppose that* $K_n < \mu_{n+1}$. *Then there exists an eigenvalue* λ_n *of the inequality* (1) *such that*

$$\mu_n \leqq \lambda_n \leqq K_n.$$

If the eigenspace to μ_n *does not belong to the cone* V, *then there exists an eigenvalue* λ_n *with* $\mu_n < \lambda_n \leqq K_n$.

The proof is based on topological techniques due to Krasnosel'skiĭ [12] and on a construction of admissible sets in a manner similar to Theorem 1.

The eigenvalues λ_n of Theorem 4 are also points of bifurcation for associated nonlinear problems of the type (2) (see [19]).

One can find in Kučera [15] interesting results to the higher eigenvalue problem.

The case when V is a symmetric set (that is, for any $v \in V$, $-v$ also belongs to V) was treated in Dias [6] and Dias, Hernandez [7] by using a variant of the Ljusternik-Schnirelman cathegory method. See also Beirão da Veiga, Dias [2].

Unfortunately, the meaning of higher eigenvalues from a physical point of view does not seem to be clear up to the present time.

EXAMPLE. In order to illustrate Theorem 4 we consider a problem for the clamped circular plate [2?]. Let

$$\Omega_R = \{x \mid x_1^2 + x_2^2 \leqq R^2\} , \qquad 0 < R < \infty ,$$

and V the cone

$$V = \{v \in \overset{\circ}{H}_{2,2}(\Omega_R) \mid v(x) \geqq 0 \quad \text{for} \quad x \in A, \quad v(x) \leqq 0 \quad \text{for} \quad x \in B\}.$$

By A, B we denote subsets of the annulus

$$\{x \mid R_1 \leqq \sqrt{x_1^2 + x_2^2} \leqq R\} , \qquad 0 < R_1 < R.$$

The compressive forces are acting in the inner normal direction. The inequality modeling this problem is given by

$$u \in V: \int_{\Omega_R} \Delta u \Delta(v - u)\, dx \geqq \lambda \int_{\Omega_R} \{u_{x_1}(v - u)_{x_1} + u_{x_2}(v - u)_{x_2}\}\, dx$$

$$\text{for all} \quad v \in V.$$

Let τ_n be the zeros of the Bessel functions $J_\mu(x)$, $\mu = 1,2,\ldots$ which are ordered according to their magnitudes $\tau_1 < \tau_2 < \cdots$ ($\tau_1 = 3.832$, $\tau_2 = 5.136$, $\tau_3 = 6.380$, ...).

If

$$\frac{\tau_n}{\tau_{n+1}} < \frac{R_1}{R} ,$$

then there exists an eigenvalue λ_n *of the above variational inequality* with

$$\frac{\tau_n^2}{R^2} \leqq \lambda_n \leqq \frac{\tau_n^2}{R_1^2} .$$

IV. EIGENVALUE EQUATIONS IN CONVEX CONES. There exist a lot of papers on this subject, for example, Kreĭn, Rutman [14], Krasnosel'skiĭ [13], Cohen [5], to name a few.

Let $K \subset H$ be a convex closed cone with vertex at zero. We look for nonzero solutions $u \in K$ of the equation

$$a(u,v) = \lambda b(u,v) \qquad \text{for all} \quad v \in H. \tag{6}$$

For the bounded, symmetric and real bilinear forms $a(u,v)$, $b(u,v)$ we assume that

$a(v,v)$ is coercive on H (that is, there exists $c > 0$ such that $a(v,v) \geqq c\,\|v\|^2$) for all $v \in H$,

the form $b(u,v)$ is completely continuous on H,

and there exists a vector $w \in K$ such that $b(w,w) > 0$.

It follows from Theorem 1 that λ_K defined by

$$\lambda_K^{-1} = \max_{\substack{v\in K\\ v\neq 0}} \frac{b(v,v)}{a(v,v)}$$

is the smallest positive eigenvalue of the variational inequality (1). It is well-known that λ_H defined by

$$\lambda_H^{-1} = \max_{\substack{v\in K\\ v\neq 0}} \frac{b(v,v)}{a(v,v)}$$

is the smallest positive eigenvalue of the equation (6). Let $a(u,v)$ be the sum of two bounded, symmetric and real bilinear forms

$$a(u,v) = a_1(u,v) + a_2(u,v).$$

We assume that $a_1(u,v)$ is coercive on H. Now, let us define the polar cone K' with respect to K and the inner product $a_1(u,v)$:

$$K' = \{w \in H \mid a_1(w,v) \leqq 0 \text{ for all } v \in K\}.$$

THEOREM 5 [22]. *Suppose that* (a) $u \in K'$ *implies that* $-u \in K$. (b) *For all* $v \in K$ *and for all* $w \in K'$ *we have* $a_2(v,w) \geqq 0$ *and* $b(v,w) \leqq 0$. *Then* $\lambda_H = \lambda_K$, *that is, there exists a nonzero solution* $u \in K$ *of the equation with* λ_H *as the eigenvalue.*

We have $0 < \lambda_H \leqq \lambda_K$. Let us show that $\lambda_H \geqq \lambda_K$. Let u_H be an eigenvector to λ_H. According to a theorem of Moreau [26] we can write

$$u_H = u_1 + u_2, \text{ where } u_1 \in K,\ u_2 \in K \text{ and } a_1(u_1,u_2) = 0.$$

By substituting $u = u_H$, $\lambda = \lambda_H$ and $v = u_1$ into the equation (6) we obtain from assumption (b)

$$a(u_1,u_1) \leqq \lambda_H b(u_1,u_1).$$

If $u_1 \neq 0$, then $\lambda_H \geqq \lambda_K$. In the case $u_1 = 0$ we have $u_H \in K$ and therefore $-u_H \in K$ by assumption (a).

Under the assumptions of Theorem 5 we have

THEOREM 6 [22]. *Suppose* $a(u_1,u_2) \neq 0$ *for any two eigenvectors* $u_1,u_2 \in K$ *of the equation* (6). *Then* λ_H *is a simple eigenvalue.*

EXAMPLE. Let $\Omega \subset \mathbb{R}^2$ be a simple connected, bounded and sufficiently regular domain. We consider the boundary value problems

$$\begin{aligned} \Delta^2 u &= -\lambda\Delta u \quad \text{in } \Omega \\ u &= 0, \quad Mu = 0 \quad \text{on } \partial\Omega \end{aligned} \tag{7}$$

and

$$\begin{aligned} \Delta^2 u &= \lambda u \quad \text{in } \Omega \\ u &= 0, \quad Mu = 0 \quad \text{on } \partial\Omega, \end{aligned} \tag{8}$$

where

$$Mu = \Delta u + (1 - \nu)(2 \sin\theta \cos\theta \, u_{x_1x_2} - \sin^2\theta \, u_{x_1x_1} - \cos^2\theta \, u_{x_2x_2}).$$

Here θ denotes the angle between the x_1-axis and the outer normal at $\partial\Omega$ (see Figure 3). By ν we denote the Poisson ratio $0 < \nu < 1$.

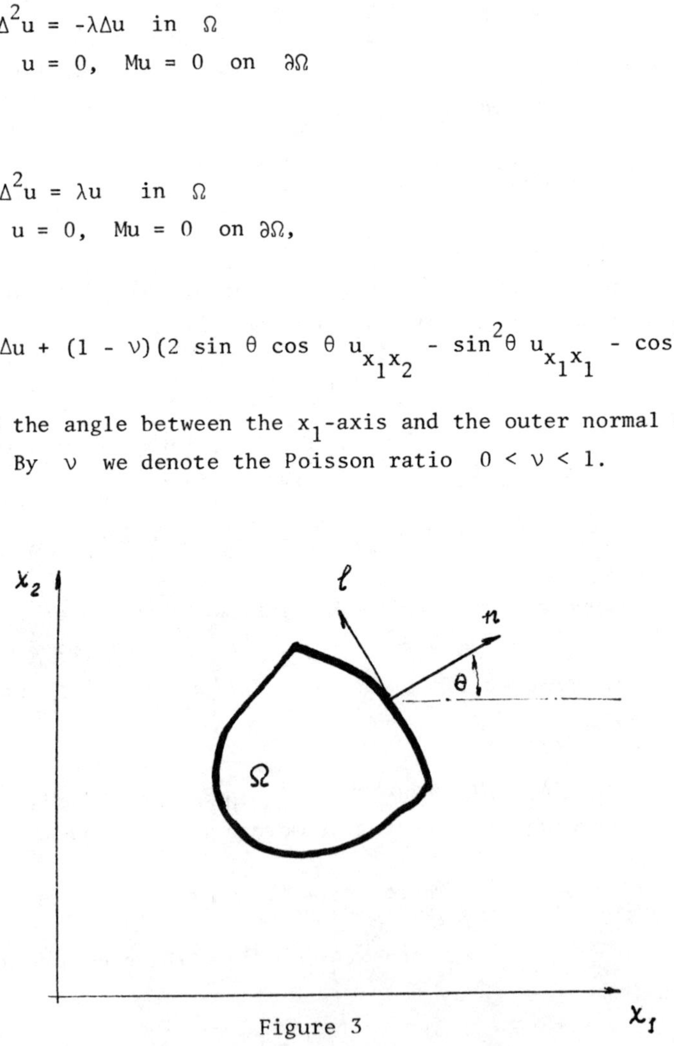

Figure 3

The equation (7) describes a buckling problem for the simply supported plate and (8) arises from the vibration problem for a simply supported plate.

The following result was proved in [22].

> Suppose Ω is convex. Then there exists a positive eigenfunction of (7) associated with the lowest eigenvalue, which is simple.

The same result is true for (8) and also for convex and bounded domains with piecewise smooth boundaries.

The weak form of equation (7) is given by (see [22])

$$u \in H: \int_\Omega \Delta u \Delta v \, dx - (1-\nu) \int_{\partial\Omega} \frac{\partial\theta}{\partial\ell} \frac{\partial u}{\partial n} \frac{\partial v}{\partial n} \, ds = \lambda \int_\Omega u_{x_i} v_{x_i} \, dx,$$

$$\text{for all } v \in H,$$

where $H = \overset{\circ}{H}_{1,2}(\Omega) \cap H_{2,2}(\Omega)$.

For the illustration of our notation see Figure 3.

Now, let us set

$$a_1(u,v) = \int_\Omega \Delta u \Delta v \, dx$$

$$a_2(u,v) = -(1-\nu)\int_{\partial\Omega} \frac{\partial\theta}{\partial\ell} \frac{\partial u}{\partial n} \frac{\partial v}{\partial n} \, ds,$$

$$b(u,v) = \int_\partial u_{x_i} v_{x_i} \, dx.$$

It is not hard to see that all assumptions of the Theorems 5 and 6 are fulfilled.

V. STABILITY PROBLEMS. We suppose that the bilinear form $a(u,v)$ is coercive on H and $V \subset H$ is convex and closed, and that $0 \in V$. Set

$$I_\lambda(v) = \frac{1}{2} a(v,v) - \frac{\lambda}{2} b(v,v). \tag{9}$$

A vector (λ_o, u_o), $u_o \in V$, $\lambda_o \in \mathbb{R}$, is said to be a strong local minimum of (9) if there exists positive numbers ρ, c such that

$$I_{\lambda_o}(v) - I_{\lambda_o}(u_o) \geqq c\,\|v - u_o\|^2 \quad \text{for all } v \in V,$$

if $\|v - u_o\| \geqq \rho$. The constant c does not depend on v.

Local extrema in connection with nonlinear variational equations were studied by Beckert [1].

Let (λ, u) be a solution of the variational inequality (1). A criterion will be given for (λ, u) to define a strong local minimum of the functional (9). For $t > 0$ we set

$$V_t(u) = \{w \in H \mid a(w,w) = 1, \quad u + tw \in V\}.$$

We denote by $K_{\lambda,u}$ the closure of the set

$$\{h = t(v - u) \mid t > 0, \quad F_{\lambda,u}(v - u) = 0, \quad v \in V\},$$

where

$$F_{\lambda,u}(w) = a(u,w) - \lambda v(u,w).$$

We assume $K_{\lambda,u} \neq \{0\}$ and for $\mu_{\lambda,u}$ defined by

$$\mu_{\lambda,u}^{-1} = \max_{\substack{h \in K_{\lambda,u} \\ h \neq 0}} \frac{b(h,h)}{a(h,h)}$$

the inequality $\mu_{\lambda,u} \quad 0.$

HYPOTHESIS H_o. For every sequence $t_n \to 0$, $t_n > 0$, and for every weakly convergent sequence $w_n \rightharpoonup w$, $w_n \in V_{t_n}(u)$

$$\lim_{n\to\infty} \frac{F_{\lambda,u}(w_u)}{t_n} < \infty$$

implies the inequality

$$1 - \lambda b(w,w) > 0$$

THEOREM 7 [24]. *Under the hypothesis* H_o *a solution* (λ,u) *of the variational inequality* (1) *defines a strong local minimum of the functional* (9).

Now let (λ,u_λ), $\lambda_1 < \lambda < \lambda_2$, $\lambda_1 < \lambda_2$, be a continuous branch $\mathcal{L}$ of solutions to (1). A branch is said to be continuous if $u_\lambda \to u_{\lambda_o}$, where $\lambda,\lambda_o \in (\lambda_1,\lambda_2)$.

HYPOTHESIS H_1. Let $(\lambda_n,u_n) \in \mathcal{L}$ be a sequence where $\lambda_n \to \lambda_o$. For every sequence $t_n \to 0$, $t_n > 0$, and for every weakly convergent sequence $w_n \rightharpoonup w$, $w_n \in V_{t_n}(u_n)$ the inequality

$$1 - \lambda_o b(w,w) > 0$$

follows from

$$\lim_{n\to\infty} \frac{F_{\lambda_n,u_n}(w_n)}{t_n} < \infty .$$

THEOREM 8 [25]. *Under the assumption* $\mu^{-1}_{\lambda,u} \geqq c > 0$ *where* c *does not depend on* $\lambda \in (\lambda_1, \lambda_2)$, *and under the hypothesis* H_1 *there is no bifurcation from* $\mathcal{L}$. *That means there does not exist a sequence* $(\lambda_n, u_n) \notin \mathcal{L}$ *of solutions to* (1) *such that* $\lambda_n \to \lambda$, $\lambda_1\lambda_n \in (\lambda_1, \lambda_2)$, *and* $u_n \to u_\lambda$.

The above results can be generalized to the nonlinear inequality (2) with respect to the functional

$$I_\lambda(v) = f(v) - \lambda g(v)$$

(see [24]).

EXAMPLE. Applications to plate theory were studied in [25] and to beam theory in [24]. For simplicity, we consider the beam theory.

The energy of the compressed beam (in the linear theory) is given by

$$I_\lambda(v) = \frac{1}{2} EJ \left[\int_0^\ell v''^2 \, dx - \lambda \int_0^\ell v'^2 \, dx\right]$$

where $\lambda = P/EJ$, ℓ is the length of the beam, EJ - bending stiffness. Suppose that the beam is simply supported at the ends, that is, the boundary conditions $v(0) = v(\ell) = 0$ are prescribed. We denote

$$H = \overset{\circ}{H}_{1,2}(0,\ell) \cap H_{2,2}(0,\ell)$$

and

$$V = \{v \in H \mid |v(x)| \leqq d \text{ on } (0,\ell)\},$$

where $0 < d = \text{const}$.

The family of functions

$$u_\lambda = \begin{cases} \frac{d}{\pi}(\sqrt{\lambda}x + \sin\sqrt{\lambda}x) & \text{if } 0 \leqq x \leqq k \\ d & \text{if } k < x < \ell - k \\ \frac{d}{\pi}(\sqrt{\lambda}(\ell - x) + \sin\sqrt{\lambda}(\ell - x) & \text{if } \ell - k \leqq x \leqq \ell , \end{cases}$$

where $\lambda = (\pi/k)^2$ and $0 < k \leq 1/2$, define solutions to the variational inequality

$$u \in V: \quad \int_0^{\pi} u''(v-u)''\,dx \geq \lambda \int_0^{\ell} u'(v-u)'\,dx \quad \text{for all} \quad v \in V.$$

It was proved in [24] that

i) The above solution (λ, u_λ) is a strong local minimum of the functional (10) provided

$$\left(\frac{2\pi}{\ell}\right)^2 \leq \lambda < \left(\frac{4\pi}{\ell}\right)^2 .$$

ii) There is no bifurcation from the branch if

$$\left(\frac{2\pi}{\ell}\right)^2 \leq \lambda < \left(\frac{4\pi}{\ell}\right)^2 .$$

iii) The solution $(\lambda_o, u_{\lambda_o})$ of the family is a point of bifurcation where $\lambda_o = (4\pi/\ell)^2$.

The proof is based on the Theorems 7, 8, and on Taylor's theorem for functions belonging to Sobolev spaces. We shall sketch the proof of the assertion i).

After integration by parts we obtain

$$F_{\lambda,u}(w) = -\frac{\pi^2 d}{k^3}\,[w(k) + w\,(\ell - k)].$$

If $F_{\lambda,u}(w_n) = 0$ where $w_n \rightharpoonup w$, $w_n \in V_{t_n}(u)$, $t_n \to 0$, then it follows that $w_n(k) = w_n(\ell - k) = 0$. In order to prove $w_n'(k) = w_n'(\ell - k) = 0$ we use Taylor's theorem. Since

$$u + t_n w_n \leq d,$$

we deduce that

$$t_n(w_n'(k)h - c_1|h|^{3/2}) \leq c_2|h|^{3/2},$$

for all h, $|h| < h_o$, with h_o sufficiently small. Therefore, we have $w_n'(k) = 0$. By c_k we denote positive constants which do not depend on n and h. In the same manner we get $w_n'(\ell - k) = 0$.

Now let us assume that for a weakly convergent subsequence $w_n \rightharpoonup w$, $w_n \in V_{t_n}(u)$, $t_n \to 0$, we have $w_n(k) < 0$. Since $F_{\lambda,u}(w) = 0$ it follows that $w(k) = 0$. The equation $w'(k) = 0$ we deduce again from Taylor's theorem. We have $u \in H_{3,2}(0,1)$. It can be shown (Frehse [10]) that every solution u of the above variational inequality belongs to the class $H_{3,2}(0,1)$. By using Taylor's theorem we get

$$t_n(w_n(k) + w'(k)h - c_3|h|^{3/2}) \leqq c_4|h|^{5/2}.$$

Suppose that $w_n'(k) \leqq -\alpha$ for an $\alpha > 0$ and for all $n \geqq n_o$. For $h < 0$ we obtain

$$t_n(w_n(k) - \alpha h - c_3|h|^{3/2}) \leqq c_4|h|^{5/2}$$

Putting

$$h = 2\,\frac{w_n(k)}{\alpha}$$

and using the abbreviation $a_n = |w_n(k)|$, from the last inequality that

$$t_n \leqq \frac{c_5 c_5 a_n^{5/2}}{a_n - c_6 a_n^{3/2}},$$

and therefore

$$\frac{F_{\lambda,u}(w_n)}{t_n} \geqq c_7(a_n^{-1/2} - c_8) \to +\infty \quad \text{as} \quad n \to \infty$$

since a_n tends to zero. The inequalities $w'(k) \leqq 0$ and $w'(\ell - k) = 0$ can be proved by the use of identical techniques.

REFERENCES

[1] H. Bechert, Variations - und Eigenwertprobleme zu nichtlinearen Differentialgleichungssystemen höherer Ordnung, Math. Machr. 49(1971), 311-341.

[2] H. Beirão da Veiga and J.-P. Dias, Sur la surjectivité de certains opérateurs non linéaires liés aux inéquations variationnelles, Bull. Un. Mat. Ital. 10(1974), 52-59.

[3] N, Berger, On von Karman's equations and the buckling of a thin elastic plate, Comm. Pure Appl. Math., 20(1967), 687-719.

[4] M. Berger and P. Fife, Von Karman's equations and the buckling of a thin elastic plate. II. Plate with edge conditions, Comm. Pure Appl. Math. 21(1968), 227-241.

[5] D. S. Cohen, Positive solutions of a class of nonlinear eigenvalue problems, J. Math. Mech., 17(1967), 209-215.

[6] J.-P. Dias, Variational inequalities and eigenvalue problems for nonlinear maximal monotone operators in a Hilbert space, Amer. J. Math., 97(1975), 905-914.

[7] J.-P. Dias and J. Hernandez, A Sturm -Liouville theorem for some odd multivalued maps, Proc. Amer. Math. Soc., 53(1975), 72-74.

[8] Cl. Do, Problèmes de valeurs propres pour une inéquation variationnelle sur un cône et application au flembement unilatéral d'une plaque mince, C. R. Acad. Sci. Paris Ser A-B, 280(1975), 45-48.

[9] _____, Bifurcation theory for elastic plates subjected to unilateral conditions, J. Math. Anal. Appl., 60(1977), 435-448.

[10] J. Frehse, Zum Differenzierbarkeitsproblem bei Variationsungleichungen höherer Ordnung, Abh. Math. Sem. Univ. Hamburg, 36(1971), 140-149.

[11] Th. von Kármán, Festigkeitsprobleme im Maschinenbau, Encyklopädie der Math. Wissensch. Band IV, 311-385 (1910).

[12] M. A. Krasnosel'skiǐ, Topological methods in the theory of nonlinear integral equations, in Russian, Moscow, 1956.

[13] _____, Positive solutions of operator equations, in Russian, Moscow, 1962.

[14] M. G. Kreǐn and M. A. Rutman, Linear operators leaving a cone invariant in a Banach space, Uspehi Mat. Nauk, 3(1948), 3-95.

[15] M. Kučera, A new method for obtaining eigenvalues of variational inequalities based on bifurcation theory, Casopis Pest. Math., 104(1979), 389-411.

[16] H. Link, Über den geraden Knickstab mit begrenzter Durchbiegung, Ing.-Arch., 22(1954), 237-250.

[17] J. L. Lions and G. Stampacchia, Variational inequalities, Comm. Pure Appl. Math., 20(1967), 493-519.

[18] E. Miersemann, Verzweigungsprobleme für Variationsungleichungen, Math. Nachr., 65(1975), 187-209.

[19] _____, Über höhere Verzweigungspunkte nichtlinearer Variationsungleichungen, Math. Nachr., 85(1978), 195-213.

[20] _____, Über nichtlineare Eigenwertaufgaben in konvexen Mengen, Math. Nachr., 88(1979), 191-205.

[21] _____, Ein Beulproblem fur die eingespannte Kreisplatte mit Unterstutzungen, Z. Angew. Math. Mech., 57(1977), 717.

[22] _____, Über positive Lösungen von Eigenwertgleichungen mit Anwendungen auf elliptische Gleichungen zweiter Ordnung und auf ein Beulproblem für die Platte, Z. Angew. Math. Mech., 59(1979), 189-194.

[23] _____, Eigenwertaufgaben für Variationsungleichungen, Math. Nachr., to appear.

[24] _____, Zur Lösungsverzweigung bei Variationsungleichungen mit einer Anwendung auf den Knickstab mit begrenzter Durchbiegung, Math. Nachr., to appear.

[25] _____, Stabilitatsprobleme für Eigenwertaufgaben bei Beschränkungen für die Variationen mit einer Anwendung auf die Platte, Math. Nachr., to appear.

[26] J. J. Moreau, Décomposition orthogonale dans un espace Hilbertien selon deux cones mutuellement polaires, C. R. Acad. Sci. Paris Sér A-B, 255(1962), 238-240.

[27] E. Trefftz, Zur Stabilität des elastischen Gleichgewichts, Z. Angew. Math. Mech., 13(1933), 160-165.

ERICH MIERSEMANN
KARL-MARX-UNIVERSITÄT
SEKTION MATHEMATIK
DDR-7010 LEIPZIG
GERMAN DEMOCRATIC REPUBLIC

R.R.W.

Contemporary Mathematics
Volume 4
1981

REGULARITY AND SYMMETRY PROPERTIES OF SOLUTIONS OF THE JOHN SHELL EQUATIONS FOR A SPHERICAL SHELL

George H. Knightly[1] and D. Sather[2]

ABSTRACT. A method is outlined to obtain classical buckled states of a (complete) thin, elastic spherical shell under uniform external pressure. The shell model investigated leads to the John shell equations, a system of six nonlinear partial differential equations. The idea of weak solutions together with methods of bifurcation theory and group representations are used to obtain classical buckled states with various symmetry and stability properties.

One of the most perplexing questions facing an investigator in nonlinear shell theory is what model (or set of equations) to use in the study of a particular problem. The approximate equations obtained from a theory of shallow shells (e.g., see [6]) are usually not appropriate for the study of "global" (or "large") solutions whereas the highly nonlinear equations obtained from a geometrically exact theory of shells (e.g., see [1]) are usually too complex to permit a complete analysis of the problem. Thus, the investigator is often forced to make some sort of compromise between the classical and geometrically exact theories of thin, elastic shells.

Such a compromise is offered here through the use of the John shell equations to describe the buckled states of a thin, elastic spherical shell subjected to a uniform external pressure. The John shell equations, a system of six coupled nonlinear partial differential equations, were derived in [2] by means of rigorous a priori estimates of elliptic partial differential equations and in [6,§9] by means of a formal variational principle. The approach in [2] yields both an approximate set of two-dimensional shell equations and rigorous estimates of the "higher order" error terms. The John equations are derived under the assumption that the strains (or stresses) are sufficiently small but there are no a priori assumptions made on the size of the displacements of the shell. The (intrinsic) equations are formulated in terms of the first and second fundamental forms of the deformed "middle" surface of the shell; once the first and second fundamental forms have been determined the displacements

1980 Mathematics Subject Classification. 35J65, 47H17, 73H05, 73K15
[1]Supported in part by NSF Grant No. MCS79-03555
[2]Supported in part by NSF Grant No. MCS79-23650

of the middle surface of the shell are obtained by integrating a system of ordinary differential equations. In terms of a tensor of stress resultants $n_{\alpha\beta}$ and a tensor of curvature changes $\rho_{\alpha\beta}$ the John shell equations may be written as

$$(1.1)\quad \begin{cases} \frac{1}{Eh} n^{\alpha}_{\alpha}|^{\beta}_{\beta} + \tilde{b}^{\alpha}_{\alpha}\rho^{\beta}_{\beta} - \tilde{b}^{\alpha}_{\beta}\rho^{\beta}_{\alpha} + \frac{1}{2}(\rho^{\alpha}_{\alpha}\rho^{\beta}_{\beta} - \rho^{\alpha}_{\beta}\rho^{\beta}_{\alpha}) = 0 \\ \rho^{\alpha}_{\beta}|_{\mu} - \rho^{\alpha}_{\mu}|_{\beta} = 0 \\ \frac{Eh^3}{12(1-\nu^2)}\rho^{\alpha}_{\alpha}|^{\beta}_{\beta} - (\tilde{b}^{\alpha}_{\beta} + \rho^{\alpha}_{\beta})n^{\beta}_{\alpha} = -p \\ n^{\alpha}_{\beta}|_{\alpha} = 0. \end{cases}$$

Here $n_{\alpha\beta}$ depends linearly on the strain tensor $\gamma_{\alpha\beta}$ representing the difference between the first fundamental (metric) tensors $\tilde{a}_{\alpha\beta}$ and $\bar{a}_{\alpha\beta}$ of the undeformed and deformed middle surfaces of the shell, respectively, and $\rho_{\alpha\beta}$ represents the difference between the second fundamental tensors $\tilde{b}_{\alpha\beta}$ and $\bar{b}_{\alpha\beta}$ of the undeformed and deformed middle surfaces of the shell, respectively. The pressure p is positive when it is directed inward, h denotes the uniform thickness of the shell, E denotes Young's modulus, and ν denotes Poisson's ratio. The raising and lowering of indices is performed relative to $\tilde{a}_{\alpha\beta}$, the symbol "|" followed by additional subscripts denotes covariant differentiation relative to $\tilde{a}_{\alpha\beta}$, and a comma followed by an additional subscript denotes partial differentiation. The usual summation convention is used with Greek indices ranging over the values 1 and 2. As formulated in (1.1), the John shell equations offer the advantages of a model with a rational basis for study, and a model that provides accurate results when the strains are "small" even if the displacements are "large".

A class of generalized solutions of the John shell equations for a spherical shell under uniform external pressure was introduced in [3] and studied further in [4]. Among other things the existence of generalized solutions with various symmetry and stability properties was established in [3] for loads λ near the classical buckling load λ_1. As pointed out in [3], however, the generalized solutions obtained there were not classical solutions because constants were not admitted as test functions. (The same kind of difficulty arises when constants are ignored in obtaining formal solutions of the buckling problem for a spherical shell as in [7]). The generalized solutions found in [3] do, however, correspond to classical solutions of the buckling problem. In fact, in this paper we show, among other things, that each generalized solution obtained in [3] corresponds to a classical solution having the same leading term and the same symmetry and stability properties.

In order to establish the desired correspondence between classical solutions and the generalized solutions obtained in [3], we consider first a more restrictive but more natural class of generalized solutions, which we shall call weak solutions in order to distinguish them from the generalized solutions in [3]. With the definition of weak solution given here, it is proved that a solution of the John equations is a weak solution if and only if it is a classical solution. We then show that the appropriate solvability conditions for the buckling problem (i.e., the reduced bifurcation equations) are the same for both weak and generalized solutions. Thus, in the context of "small" solutions near $\lambda = \lambda_1$ the problem of determining the generalized solutions obtained in [3] is equivalent to determining the weak (hence classical) solutions obtained in the present paper. In particular, these results show that the generalized solutions in [3] and the formal solutions in [7] always generate classical solutions of their respective problems.

For the problem of a spherical shell, it is natural to obtain solutions with special symmetry properties by making use of the invariance of the problem under subgroups of the orthogonal group $\mathcal{O}(3)$. Such an application of group representations is greatly simplified when a variational structure enables one to obtain nontrivial solutions of the problem as the nontrivial critical points of a smooth potential functional. In the present paper we derive such a potential functional and then use unitary representations of $\mathcal{O}(3)$ to carry out the necessary symmetry arguments.

The use of energy methods to determine the possible buckled states of a thin, elastic shell is well-known in the nonlinear theory of shallow shells. In particular, the work of Koiter in [5;6;7] is based upon energy considerations and has led to a "general nonlinear theory of stability". The reader is referred to [7] and to [3;4] for an extensive introduction to the problem of the buckling of a spherical shell.

2. CLASSICAL AND WEAK SOLUTIONS. Let $\mathcal{S}_R$ and $\mathcal{S}$ denote the spheres in $\mathbb{R}^3$ of radius R and 1, respectively, with centers at the origin of a Cartesian coordinate system (x_1,x_2,x_3). Let (r,θ,ϕ) be a system of spherical coordinates with polar angle θ measured from the x_3-axis. We assume that the middle surface of the undeformed (complete) spherical shell coincides with $\mathcal{S}_R$ and is described by $r = R$, $0 \le \theta \le \pi$ and $0 \le \phi \le 2\pi$. The covariant metric tensor $\tilde{a}_{\alpha\beta}$ on $\mathcal{S}_R$ is given by

$$\tilde{a}_{\alpha\beta} = \begin{pmatrix} R^2 & 0 \\ 0 & R^2 \sin^2 \theta \end{pmatrix} \tag{2.1}$$

and the metric on $\mathcal{S}_R$ is given by $ds^2 = R^2 d\theta^2 + R^2 \sin^2 \theta d\phi^2$. We denote the contravariant metric tensor on $\mathcal{S}_R$ by $\tilde{a}^{\alpha\beta}$ and the usual alternating tensors by $\tilde{\varepsilon}_{\alpha\beta}$ and $\tilde{\varepsilon}^{\alpha\beta}$, i.e.,

$$\tilde{\varepsilon}_{\alpha\beta} = \begin{pmatrix} 0 & \tilde{a}^{1/2} \\ -\tilde{a}^{1/2} & 0 \end{pmatrix}, \tag{2.2}$$

where $\tilde{a} = \det \tilde{a}_{\alpha\beta}$. The corresponding tensors on $\mathcal{S}$ will be denoted by $a_{\alpha\beta}$, $a^{\alpha\beta}$, $\varepsilon_{\alpha\beta}$ and $\varepsilon^{\alpha\beta}$.

In the case of spherical shells, for which

$$\tilde{b}_{\alpha\beta} = -\frac{1}{R}\tilde{a}_{\alpha\beta} \tag{2.3}$$

we shall show that the John equations in (1.1) have a simple formulation as a single operator equation in a Hilbert space. First, it is useful to show that the system of equations in (1.1) can be replaced by a coupled pair of nonlinear partial differential equations for a stress function $\tilde{f}$ and a generalized curvature function $\tilde{w}$, namely

$$\Delta^2\tilde{f} + 2\Delta\tilde{f} - \alpha(\Delta\tilde{w} + 2\tilde{w}) + \frac{1}{2}\{\tilde{w},\tilde{w}\} = 0 \tag{2.4a}$$

$$\Delta^2\tilde{w} + 2\Delta\tilde{w} + \alpha(\Delta\tilde{f} + 2\tilde{f}) - \{\tilde{f},\tilde{w}\} = -2\alpha\lambda . \tag{2.4b}$$

Here $\gamma = [12(1 - \nu^2)]^{-1}$,

$$\alpha = (\gamma h)^{-1}R \text{ and } \lambda = \frac{pR^3}{2Eh^3\gamma^2} , \tag{2.5}$$

$\Delta w \equiv a^{\alpha\beta}w|_{\alpha\beta}$ denotes the Laplace-Beltrami operator on $\mathcal{S}$, and

$$\{u,v\} = (\varepsilon^{\sigma\kappa}\varepsilon^{\beta\mu}u_{|\kappa\mu}v_{,\sigma})_{|\beta} + a^{\sigma\mu}u_{,\sigma}v_{,\mu} + u\Delta v + v\Delta u + 2uv . \tag{2.6}$$

It was shown in [3] that the spherical harmonics (suitably normalized) form a natural basis for the problem. Since the spherical harmonics of degree zero and degree one span the null space of the linear operator $\Delta^2 + 2\Delta$ appearing in (2.4), it is natural in (2.4) to set $\tilde{f} = f^0 + f^1 + f$ and $\tilde{w} = w^0 + w^1 + w$, where f^0 and w^0 are spherical harmonics of degree zero (i.e., constants), f^1 and w^1 are spherical harmonics of degree one, and f and w are functions to be determined. If one then equates the terms not containing f and w, one obtains the equations

$$\begin{cases} -2\alpha w^0 + (w^0)^2 = 0 , \\ 2\alpha f^0 - 2f^0 w^0 = -2\alpha\lambda . \end{cases} \tag{2.7}$$

There are two solutions to this system, namely $(f^0,w^0) = (-\lambda,0)$ and $(f^0,w^0) = (\lambda,2\alpha)$. We choose the first of these solutions since it reduces to $(0,0)$ when the load λ is zero. With this choice of f^0 and w^0 the system (2.4) becomes

$$\Delta^2 f + 2\Delta f - \alpha(\Delta w + 2w) + \frac{1}{2}\{w,w\} = 0, \tag{2.8a}$$

$$\Delta^2 w + 2\Delta w + \alpha(\Delta f + 2f) + \lambda(\Delta w + 2w) - \{f,w\} = 0. \tag{2.8b}$$

Note that the terms from (2.4) containing f^1 and w^1 cancel in (2.8). Since the contributions of f^1 and w^1 to the stresses and curvature changes are also zero (see (A.5) in Appendix A of [3]), it follows that the spherical harmonics of degree one play no role in the model being used here. We shall see, however, that the spherical harmonics of degree zero play a crucial role in solving (2.8).

The problem of determining a solution pair w,f of (2.8) shall be called Problem S. A <u>classical solution of Problem S</u> is a pair of functions w,f that are four times continuously differentiable on $\mathcal{S}$ and satisfy equations (2.8) pointwise on $\mathcal{S}$.

In order to define a weak solution of (2.8) we first introduce the (real) Sobolev space $W_{2,2}(\mathcal{S})$ defined as the closure of smooth functions on $\mathcal{S}$ in the usual norm $\|\cdot\|_{2,2}$. We take $\mathcal{H}$ to be the closed linear subspace of $W_{2,2}(\mathcal{S})$ consisting of elements in $W_{2,2}(\mathcal{S})$ that are orthogonal, in the usual L^2 inner product on $\mathcal{S}$, to $P_0(\cos\theta) \equiv 1$, $P_1(\cos\theta)$, $P_1^1(\cos\theta)\cos\phi$ and $P_1^1(\cos\theta)\sin\phi$, where P_n denotes the Legendre polynomial of degree n and P_n^k denotes the associated Legendre function of the first kind of order k (e.g., see [11,p.127]). That is, $w \in \mathcal{H}$ if and only if $w \in W_{2,2}(\mathcal{S})$ and

$$\int w\,dS = \int wP_1(\cos\theta)dS = \int wP_1^1(\cos\theta)\cos\phi\, dS = \int wP_1^1(\cos\theta)\sin\phi\, dS = 0,$$

where $dS = \sin\theta\, d\theta\, d\phi$ is the element of area on $\mathcal{S}$. (Integration over $\mathcal{S}$ will be denoted throughout the paper by an integral with no limits.) One can now show that the bilinear form $(\cdot,\cdot)$ defined by

$$(u,v) = \int(\Delta u\Delta v + 2u\Delta v)dS, \qquad u, v \in \mathcal{H}, \tag{2.9}$$

is symmetric and coercive over $\mathcal{H}$. The bilinear form $(\cdot,\cdot)$ determines a natural inner product for $\mathcal{H}$ and is used throughout the paper; the corresponding norm on $\mathcal{H}$ is denoted by $\|\cdot\|$.

As pointed out in the above discussion, the spherical harmonics of degree zero cannot be ignored in solving the system in (2.8). To deal with these constants we introduce a second Hilbert space $\mathcal{W}$ by adjoining the constants to $\mathcal{H}$:

$$\mathcal{W} = \mathbb{R}^1 \oplus \mathcal{H} = \{u_0 + u_1 : u_0 \in \mathbb{R}^1,\ u_1 \in \mathcal{H}\}. \tag{2.10}$$

The inner product $\langle\cdot,\cdot\rangle$ and norm $\langle\langle\cdot\rangle\rangle$ on $\mathcal{W}$ are given by

$$\langle u,v\rangle = u_0v_0 + (u_1,v_1), \qquad \langle\langle u\rangle\rangle = \langle u,u\rangle^{1/2}, \tag{2.11}$$

where (u_1,v_1) is the inner product (2.9) on $\mathcal{H}$. In (2.10), (2.11) and in the remainder of the paper the subscript "o" on an element u of $\mathcal{W}$ denotes its mean value over $\mathcal{S}$ or, equivalently, its projection E_0u onto $\mathbb{R}^1$, i.e.,

(2.12) $$u_0 = E_0u = \langle u,1\rangle = \frac{1}{4\pi}\int udS.$$

Similarly, the subscript "1" on u denotes its projection E_1u onto $\mathcal{H}$, i.e., $E_1 = I - E_0$ and

(2.13) $$u_1 = E_1u \in \mathcal{H},\ \langle E_1u,v\rangle = \int(\Delta u\Delta v + 2v\Delta u)dS = (u_1,v_1),\ u,v \in \mathcal{W}.$$

Now let u,v be smooth functions in $\mathcal{W}$. If we multiply (2.8a) by u and (2.8b) by v, and integrate the resulting equations by parts over $\mathcal{S}$, we obtain

(2.14a) $$(f_1,u_1) = -\alpha a(w;u) - \tfrac{1}{2}b(w,w;u),$$

(2.14b) $$(w_1,v_1) = \lambda a(w;v) + \alpha a(f;v) + b(w,f;v),$$

where, for arbitrary $u,v,w \in \mathcal{W}$, the functionals a and b are defined by

(2.15) $$a(u;v) = \int[a^{\sigma\beta}u|_\sigma v|_\beta - 2uv]dS,$$

(2.16) $$b(u,v;w) = -\int\varepsilon^{\sigma\kappa}\varepsilon^{\beta\mu}u|_{\kappa\mu}v|_\sigma w|_\beta dS + \int[v\Delta u + u\Delta v + a^{\sigma\beta}v|_\sigma u|_\beta + 2uv]wdS.$$

DEFINITION. A weak solution of Problem S is a pair of functions w,f in $\mathcal{W}$ that satisfy (2.14) for all u,v in $\mathcal{W}$.

REMARK 2.1. The definition of weak solution given here is different from that of a generalized solution given in [3,p.326]. Since constants were not admitted as test functions in [3], the generalized solutions introduced in [3] are not classical solutions of Problem S; in this sense the generalized solutions used in [3] are only "approximate" solutions of the problem. The same kind of difficulty arises when constants are ignored in obtaining formal solutions as in [7].

If a pair w,f is a classical solution of Problem S, then so is w + u, f + v, where u and v are any spherical harmonics of degree one. Thus we may assume that w,f are orthogonal to the spherical harmonics of degree one, i.e., that $w,f \in \mathcal{W}$. The steps leading to (2.14) show that such a classical solution is also a weak solution. Conversely, the usual arguments for proving the interior regularity of weak solutions of elliptic problems apply here to show that a weak solution is a classical solution. We omit the details and state the result.

THEOREM 1. A pair of functions w,f in $\mathcal{W}$ is a classical solution of Problem S if and only if it is a weak solution of Problem S.

The remainder of this section is devoted to showing that for each generalized solution in [3] there is a classical solution with the same lowest order terms in $(\lambda - \lambda_1)$.

It is easy to extend the proof, given in [3,p.363], of the boundedness of $a(u;v)$ and $b(u,v;w)$ for $u,v,w \in \mathcal{H}$ to show that they are also bounded for $u,v,w \in \mathcal{W}$:

$$|a(u;v)| \le \text{const.} \langle\langle u\rangle\rangle \langle\langle v\rangle\rangle, \quad u,v \in \mathcal{W}, \tag{2.17}$$

$$|b(u,v;w)| \le \text{const.} \langle\langle u\rangle\rangle \langle\langle v\rangle\rangle \langle\langle w\rangle\rangle, \quad u,v,w \in \mathcal{W}. \tag{2.18}$$

Hence, by the Riesz representation theorem there are operators $A : \mathcal{W} \to \mathcal{W}$ and $B : \mathcal{W} \times \mathcal{W} \to \mathcal{W}$ such that

$$\langle Au,v\rangle = a(u;v), \quad u,v \in \mathcal{W}, \tag{2.19}$$

$$\langle B(u,v),w\rangle = b(u,v;w), \quad u,v,w \in \mathcal{W}. \tag{2.20}$$

Since $(u_1,v_1) = \langle u_1,v\rangle$ we can use the operators A and B to replace the system (2.14) by the following pair of operator equations in $\mathcal{W}$:

$$f_1 = -\alpha Aw - \tfrac{1}{2}B(w,w), \tag{2.21}$$

$$w_1 = \lambda Aw + \alpha Af + B(w,f) . \tag{2.22}$$

The next lemma summarizes some of the properties of the operators A and B.

LEMMA 1. (i) The form $\langle B(u,v),w\rangle$ is symmetric for $u,v,w \in \mathcal{W}$. In particular, $B(u,v)$ is a symmetric, bilinear operator.

(ii) The operator A is a compact, selfadjoint linear operator satisfying

$$Au = -B(u,1), \quad u \in \mathcal{W}. \tag{2.23}$$

The characteristic values $\{\mu_n\}$ of A are given by

$$\mu_0 = -(8\pi)^{-1}, \ \mu_n = n(n+1) \quad n = 2,3,\ldots . \tag{2.24}$$

The proofs of part (i) and the compactness of A follow along the lines of the proofs of Lemmas 2 and 3 in [3]. The identity (2.23) follows easily from (2.15), (2.16), (2.19), and (2.20). Since

$$\langle A1,w\rangle = \langle Aw,1\rangle = -\langle B(1,1),w\rangle = -2\int w\,dS = -8\pi E_0 w = \langle -8\pi,w\rangle$$

for all $w \in \mathcal{W}$, one sees that $A1 = -8\pi$. This last result and Lemma 2 of [3] imply (2.24). Note that the eigenspace corresponding to μ_n is $(2n+1)$-dimensional and is spanned by the spherical harmonics of degree n.

It follows from Lemma 1 and the definition of A that the spherical surface harmonics are orthogonal with respect to both of the forms $\langle u,v\rangle$ and $\langle Au,v\rangle$. Thus, $AE_0 = E_0A$ and $AE_1 = E_1A$ which imply

$$(Au)_o = Au_o = -8\pi u_o \text{ and } (Au)_1 = Au_1. \tag{2.25}$$

We may now use (2.23), (2.25) and the bilinearity of B to rewrite (2.21), (2.22) as follows:

$$f_1 + \alpha Aw_1 - w_o Aw_1 + \tfrac{1}{2}B_1(w_1,w_1) = 8\pi\alpha w_o - 4\pi w_o^2 - \tfrac{1}{2}B_o(w_1,w_1), \tag{2.26}$$

$$\begin{aligned} &w_1 - \alpha Af_1 - \lambda Aw_1 + f_o Aw_1 + w_o Af_1 - B_1(f_1,w_1) \\ &= -8\pi\alpha f_o - 8\pi\lambda w_o + 8\pi f_o w_o + B_o(f_1,w_1), \end{aligned} \tag{2.27}$$

where $B_j(u,v) = (B(u,v))_j$, $j = 0,1$. In each of the equations (2.26) and (2.27) the left side is in $\mathcal{H}$ while the right side is constant. Thus both sides are zero and we obtain a system of four equations in the four unknown f_o, f_1, w_o, and w_1, namely

$$f_1 = -\alpha Aw_1 + w_o Aw_1 - \tfrac{1}{2}B_1(w_1,w_1), \tag{2.28a}$$

$$w_1 - \alpha Af_1 - \lambda Aw_1 + f_o Aw_1 + w_o Af_1 - B_1(f_1,w_1) = 0, \tag{2.28b}$$

$$8\pi\lambda w_o + 8\pi\alpha f_o - 8\pi w_o f_o + \langle Af_1,w_1\rangle = 0, \tag{2.28c}$$

$$8\pi\alpha w_o - 4\pi w_o^2 + \tfrac{1}{2}\langle Aw_1,w_1\rangle = 0. \tag{2.28d}$$

Here in (2.28c,d) we have used

$$B_o(u,v) = \langle B(u,v),1\rangle = -\langle Au,v\rangle, \tag{2.29}$$

which follows from (2.12), (2.23) and part (i) of Lemma 1. Equation (2.28a) may now be used to eliminate f_1 from (2.28b,c,d). The resulting system can be regarded as a single equation

$$0 = \mathcal{L}_\lambda z + \mathcal{Q}(z) + \mathcal{C}(z) \tag{2.30}$$

for $z = (w_1,w_o,f_o)$ in a new Hilbert space $\mathcal{Z} = \mathcal{H} \times \mathbb{R}^2$ with inner product

$$[(w_1,w_o,f_o), (u_1,u_o,g_o)] = (w_1,u_1) + w_o u_o + f_o g_o \tag{2.31}$$

and norm $[[z]] = [z,z]^{1/2}$. In (2.30) the operators $\mathcal{L}_\lambda$, $\mathcal{Q}$ and $\mathcal{C}$ are defined by

$$\mathcal{L}_\lambda = \begin{pmatrix} L_\lambda & 0 & 0 \\ 0 & 8\pi\lambda & 8\pi\alpha \\ 0 & 8\pi\alpha & 0 \end{pmatrix} \tag{2.32}$$

$$L_\lambda = I - \lambda A + \alpha^2 A^2, \tag{2.33}$$

(2.34) $$\mathcal{Q}(w_1,w_o,f_o) = (\alpha Q(w_1) - 2\alpha w_o A^2 w_1 + f_o A w_1,$$

$$- \alpha \|Aw_1\|^2 - 8\pi f_o\, w_o,\ \tfrac{1}{2}(Aw_1,w_1) - 4\pi w_o^2),$$

(2.35) $$Q(w_1) = \tfrac{1}{2}AB_1(w_1,w_1) + B_1(Aw_1,w_1),$$

(2.36) $$\mathcal{C}(w_1,w_o,f_o) = (C(w_1) - w_o Q(w_1) + w_o^2 A^2 w_1,$$

$$- \tfrac{1}{2}(Aw_1,B_1(w_1,w_1)) + w_o \|Aw_1\|^2,\quad 0),$$

(2.37) $$C(w_1) = \tfrac{1}{2}B_1(B_1(w_1,w_1),w_1).$$

Note that the auxiliary operators L_λ, Q and C are defined and utilized only for $w_1 \in \mathcal{H}$; they are identical with the operators in [3] identified by the same symbols.

REMARK 2.2. If we solve (2.30) for $z \in \mathcal{Z}$, $z = (w_1,w_o,f_o)$, and determine f_1 by means of (2.28a), then the pair $w = w_o + w_1$, $f = f_o + f_1$ is a weak (hence classical) solution of Problem S.

Some properties of the operators in (2.30) are summarized in the next lemma. It is proved by straightforward calculations, if one makes use of Lemma 2 and Lemma 3 in [3].

LEMMA 2. (i) $\mathcal{L}_\lambda$ is bounded, linear and selfadjoint on $\mathcal{Z}$. λ_o is an eigenvalue of $\mathcal{L}_\lambda$ if and only if λ_o is an eigenvalue of L_λ; the associated eigenfunctions have the form $z = (w_1,0,0)$, where w_1 is an eigenfunction of L_{λ_o}.

(ii) The operators $\mathcal{L}_\lambda$, $\mathcal{Q}$ and $\mathcal{C}$ are homogeneous of degrees one, two and three, respectively. They are also gradient operators, i.e.,

(2.38) $$\mathcal{L}_\lambda z = \ell'(z),\ \mathcal{Q}(z) = q'(z),\ \mathcal{C}(z) = c'(z),$$

where the prime denotes the Fréchét derivative on $\mathcal{Z}$. The potentials ℓ, q and c are defined for $z = (w_1,w_o,f_o) \in \mathcal{Z}$ by

(2.39) $$\ell(z) = \tfrac{1}{2}(L_\lambda w_1,w_1) + 4\pi\lambda w_o^2 + 8\pi\alpha f_o w_o,$$

(2.40) $$q(z) = \tfrac{1}{3}\,\alpha(Q(w_1),w_1) - \alpha w_o \|Aw_1\|^2 + \tfrac{1}{2}\,f_o(Aw_1,w_1) - 4\pi f_o w_o^2,$$

(2.41) $$c(z) = \tfrac{1}{4}(C(w_1),w_1) - \tfrac{1}{3}\,w_o(Q(w_1),w_1) + \tfrac{1}{2}\,w_o^2\,\|Aw_1\|^2.$$

In particular, solutions of (2.30) are stationary points of the functional

(2.42) $$\mathcal{E}(z) = \ell(z) + q(z) + c(z).$$

It is easy to see that the unbuckled (uniformly contracted) state of the sphere corresponds to $z = 0$ for all $\lambda \in \mathbb{R}^1$. We seek buckled states bifurcating from the unbuckled state, i.e., nontrivial solutions (z,λ) of (2.30)

branching from the trivial solution $(0,\lambda)$. It is well-known that branching can occur only at the eigenvalues of $\mathcal{L}_\lambda$ and that the existence of such buckled states may be determined by the Lyapunov-Schmidt method. Thus, by part (i) of Lemma 2, it is sufficient to study the existence of branches (z,λ) near $(0,\lambda_o)$ where λ_o is an eigenvalue of L_λ.

Suppose that the null space $\mathfrak{N} \equiv \mathfrak{N}(\lambda_o) \subset \mathcal{H}$ of L_{λ_o} is k-dimensional with basis elements $v_j \in \mathcal{H}\ (j = 1,\dots,k)$ that are orthogonal in the sense that

$$(2.43)\qquad (Av_i, v_j) = \delta_{ij} \quad (i,j = 1,\dots,k).$$

Then $z_j = (v_j,0,0) \in \mathfrak{Z}\ (j = 1,\dots,k)$ spans the null space $\mathfrak{M} \equiv \mathfrak{M}(\lambda_o) \subset \mathfrak{Z}$ of $\mathcal{L}_{\lambda_o}$.

We look for a solution z of (2.30) in the form

$$(2.44)\qquad z = \sum_{j=1}^{k} \rho_j z_j + U,$$

where $\rho = (\rho_1,\dots,\rho_k) \in \mathbb{R}^k$ and U lies in $\mathfrak{M}^\perp$, the orthogonal complement of $\mathfrak{M}$ in $\mathfrak{Z}$. Equation (2.30) can now be decomposed into its projections onto $\mathfrak{M}$ and $\mathfrak{M}^\perp$. Then, for sufficiently small ρ and $\eta = \alpha^{-1}(\lambda - \lambda_o)$, say $|\rho| < r$ and $|\eta| < \eta_o$, the resulting equation on $\mathfrak{M}^\perp$ can be solved for U by means of the contraction mapping principle; the solution $U = U(\rho,\eta)$ is analytic in ρ and η. Moreover, there is a number a such that

$$(2.45)\qquad [[U(\rho,\eta)]] \le a\,|\rho|^2, \qquad |\rho| < r,\ |\eta| < \eta_o.$$

The projection of (2.30) onto $\mathfrak{M}$ then yields the <u>branching equations</u> or <u>bifurcation equations</u> associated with (2.30), namely

$$(2.46)\ 0 = -\eta\rho_i + \alpha^{-1}[\mathfrak{A}(\sum_{j=1}^{k} \rho_j z_j + U) + \mathfrak{B}(\sum_{j=1}^{k} \rho_j z_j + U), z_i] \quad (i = 1,\dots,k).$$

Thus finding solutions $z = z(\lambda)$ of (2.30) in $\mathfrak{Z}$ near $(z,\lambda) = (0,\lambda_o)$ is equivalent to finding sufficiently small solutions $\rho = \rho(\eta)$ in $\mathbb{R}^k$ of the system of analytic equations (2.46).

If we set

$$(2.47)\qquad \xi = (\xi_1,\dots,\xi_k) = \eta^{-1}\rho,\ \eta = \alpha^{-1}(\lambda - \lambda_o) \ne 0$$

in (2.46), cancel a factor η^2, recall (2.45), and formally set $\eta = 0$, we obtain the <u>reduced branching equations</u> associated with (2.30), namely

$$0 = F_i(\xi) \equiv -\xi_i + \alpha^{-1}[Q(\sum_{j=1}^{k} \xi_j z_j), z_i] \tag{2.48}$$
$$= -\xi_i + (Q(\sum_{j=1}^{k} \xi_j v_j), v_i) \qquad (i = 1,\ldots,k).$$

Suppose that $\xi^* \neq 0$ is a solution of (2.48) at which the Jacobian $J = \partial(F_1,\ldots,F_k)/\partial(\xi_1,\ldots,\xi_k)$ is different from zero. Then for $|\eta|$ sufficiently small the implicit-function theorem implies that the system (2.46) has a nontrivial solution $\rho^*(\eta) = \eta(\xi^* + b(\eta))$, where b is analytic and satisfies $\lim_{\eta\to 0} b(\eta) = 0$. This solution ρ^* then generates a nontrivial solution z^* of (2.30) of the form

$$z^*(\lambda) = \alpha^{-1}(\lambda - \lambda_0)u^* + U^*, \tag{2.49}$$

where $u^* = \sum_{j=1}^{k} \xi_j^* z_j$ and U^* is analytic in λ and satisfies

$\lim_{\lambda\to\lambda_0} (\lambda - \lambda_0)^{-1}U^* = 0.$

In [3] we used an analogous Lyapunov-Schmidt development to study the equation

$$0 = L_\lambda w + \alpha Q(w) + C(w), \qquad w \in \mathcal{H}, \tag{*}$$

where L_λ, Q and C are the operators defined by (2.33), (2.35) and (2.37). The reduced branching equations obtained in [3, see (3.9)] are identical with the system (2.48) and lead to solutions of (*) of the form

$$w^*(\lambda) = \alpha^{-1}(\lambda - \lambda_0)v^* + V^*, \tag{2.50}$$

where $v^* = \sum_{j=1}^{k} \xi_j^* v_j$ and V^* is analytic in λ and satisfies

$\lim_{\lambda\to\lambda_0} (\lambda - \lambda_0)^{-1}V^* = 0.$

Note that only the reduced branching equations (2.48) are involved in the essential steps (e.g., in finding a nontrivial solution of (2.48) at which the Jacobian is not zero) leading to a solution of <u>either</u> of the equations (2.30) or (*). It follows that finding solutions of (2.30) near $\lambda = \lambda_0$ of the form (2.49) is equivalent to finding solutions of (*) near $\lambda = \lambda_0$ of the form (2.50). We have established the following result.

THEOREM 2. Suppose w^* is a solution of (*) of the form (2.50) and suppose it is generated by a solution ξ^* of (2.48) at which $\partial(F_1,\ldots,F_k)/\partial(\xi_1,\ldots,\xi_k) \neq 0$. Then there exists $\delta > 0$ such that (2.30) has a solution z^* of the form (2.49) for $|\lambda - \lambda_0| < \delta$. In fact,

$$z^* = (w^*,0,0) + W^*, \tag{2.51}$$

where $W^* = U^* - (V^*,0,0) = o(|\lambda - \lambda_o|)$ as $\lambda \to \lambda_o$. Furthermore, if $W^* = (W_1^*, w_o^*, f_o^*)$, then $w_o^* + z_1^*, f_o^* + f_1^*$ is a classical solution of Problem S for $|\lambda - \lambda_o| < \delta$, where

$$z_1^* = w^* + W_1^* = \alpha^{-1}(\lambda - \lambda_o)v^* + (V^* + W_1^*) \tag{2.52}$$

and

$$f_1^* = -(\alpha - w_o^*)Az_1^* - \tfrac{1}{2}B_1(z_1^*, z_1^*). \tag{2.53}$$

As a consequence of Theorem 2 one sees that each of the solutions of (*) constructed in [3] near $\lambda = \lambda_1$ corresponds to a classical solution of Problem S with the same lowest order terms. In the next section we show that the solutions w^* and z^* in Theorem 2 have corresponding symmetry properties.

3. SOLUTIONS WITH SYMMETRY PROPERTIES. We begin by giving a mathematical description of some symmetry properties of solutions (see also the discussion in [3,§1] and [4,§1] of some experimental results for thin, elastic spherical shells). Such a description is probably most conveniently given in terms of group representations. The restricted group-theoretic approach used here supplements the general approach in [9;10] in that, by exploiting the properties of nonlinear gradient operators, we are able to obtain a simple and direct treatment of variational problems.

Our approach is based upon the following method of generating subspaces of a real Hilbert space $\mathcal{U}$ that are invariant under a given nonlinear operator $F: \mathcal{U} \to \mathcal{U}$ (see also [8]). Suppose that $g \to S_g$ is a unitary representation of a group G on $\mathcal{U}$ and that F commutes with the representation $g \to S_g$ in the sense that

$$F(S_g u) = S_g F(u) \tag{3.1}$$

for all $g \in G$ and all $u \in \mathcal{U}$. If K is a finite subgroup of G and if $\mathcal{U}_K$ is the closed supspace of $\mathcal{U}$ defined as

$$\mathcal{U}_K = \{u \in \mathcal{U} : S_g u = u \quad \text{for every} \quad g \in K\}, \tag{3.2}$$

then it is easy to see that $\mathcal{U}_K$ is "invariant under F" in the sense that $F(\mathcal{U}_K) \subset \mathcal{U}_K$. Thus, the nonlinear operator equation $F(u) = 0$, $u \in \mathcal{U}$, can be studied on the subspace $\mathcal{U}_K$ with only the use of restrictions rather than projections. In particular, such a construction preserves the structure of F on $\mathcal{U}_K$ because no projections on $\mathcal{U}_K$ are required. By selecting various subgroups K of G it may be possible to determine solutions of $F(u) = 0$, $u \in \mathcal{U}$, with special symmetry properties.

In the problem of a spherical shell we take G to be $\mathcal{O}(3)$, the group of orthogonal transformations of $\mathbb{R}^3$ that keep the origin fixed. Let

$g \to T_g$ denote the representation of $\mathcal{O}(3)$ on $\mathcal{H}$ given by

(3.3) $$(T_g w_1)(x) = w_1(g^{-1}x) \ , \ x \in \mathcal{S} \ , \ w_1 \in \mathcal{H} \ ,$$

where g^{-1} denotes the inverse of g in $\mathcal{O}(3)$. Since the representation described by (3.3) is known to be unitary with respect to the usual inner product on $L^2(\mathcal{S})$ (e.g., see [11]), it follows that the representation $g \to T_g$ is unitary with respect to the inner product $(\cdot,\cdot)$ on $\mathcal{H}$ defined by (2.9). Moreover, since the operators A, Q and C commute with the representation $g \to T_g$ in the sense of (3.1) (see [3,Lemma 5]), we have the following result for the operators $\mathcal{L}_\lambda$, $\mathcal{Q}$ and $\mathcal{C}$ defined in (2.32), (2.34) and (2.36).

LEMMA 3. The representation $g \to R_g$ of $\mathcal{O}(3)$ on $\mathcal{Z}$ given by

(3.4) $$(R_g z)(x) = ((T_g w_1)(x), w_o, f_o) = (w_1(g^{-1}x), w_o, f_o), \ x \in \mathcal{S} \ , \ z \in \mathcal{Z} \ ,$$

is unitary with respect to the inner product $[\cdot,\cdot]$ on $\mathcal{Z}$ defined in (2.31). Moreover, the operators $\mathcal{L}_\lambda$, $\mathcal{Q}$ and $\mathcal{C}$ commute with the representation $g \to R_g$ in the sense of (3.1), e.g., $\mathcal{Q}(R_g z) = R_g \mathcal{Q}(z)$ for all $g \in \mathcal{O}(3)$ and all $z \in \mathcal{Z}$.

It follows from Lemma 3 that the solutions of (2.30) always occur in "sheets". In fact, since $R_g \tilde{z}$ is a solution of (2.30) for all $g \in \mathcal{O}(3)$ whenever $\tilde{z}$ is a solution of (2.30), we have the following result.

THEOREM 3. If $\tilde{z} \in \mathcal{Z}$ is a solution of (2.30) then $\tilde{z}$ generates either a two or three parameter family of classical solutions of Problem S.

In general $\tilde{z}$ generates a three parameter family of solutions of (2.30). If, however, $\tilde{z}$ is an axisymmetric solution that is independent of ϕ then $\tilde{z}$ generates only a two parameter family of solutions of (2.30).

Since the presence of such "sheets" of solutions in $\mathcal{Z}$ makes it difficult to apply any of the usual kind of implicit function theorems, Theorem 3 suggests that we should seek solutions of (2.30) in invariant subspaces of $\mathcal{Z}$ rather than in $\mathcal{Z}$ itself. A convenient way to generate such invariant subspaces is to use subgroups of $\mathcal{O}(3)$ and the construction indicated in (3.2).

Let h and g_m be the elements of $\mathcal{O}(3)$ defined by

(3.5) $$h = \begin{bmatrix} 1 & 0 & 0 \\ 0 & -1 & 0 \\ 0 & 0 & 1 \end{bmatrix} \text{ and } g_m = \begin{bmatrix} \cos\frac{2\pi}{m} & -\sin\frac{2\pi}{m} & 0 \\ \sin\frac{2\pi}{m} & \cos\frac{2\pi}{m} & 0 \\ 0 & 0 & 1 \end{bmatrix} \ (m = 1,2....).$$

Let K_1 denote the subgroup of $\mathcal{O}(3)$ generated by h and g_1, and let K_m denote the subgroup of $\mathcal{O}(3)$ generated by h, g_1 and g_m $(m \geq 2)$. If one now defines $\mathcal{H}_m \equiv \mathcal{H}_{K_m}$ $(m = 1,2,...)$ as in (3.2), then each $\mathcal{H}_m$ is an invariant subspace for A, Q and C because a condition of the type (3.1) holds for each of the operators A, Q and C (see [3,Lemma 5]). (The invariant subspaces $\mathcal{H}_m$ are the same as the "cosine" subspaces $\mathcal{C}_m$ introduced in [3,p.331] and used in [3] to study buckled states with $\frac{2\pi}{m}$ - symmetry in ϕ.) Thus, if one defines $\mathfrak{Z}_m = \mathcal{H}_m \times \mathbb{R}^2$ (m = 1,2,...) then Lemma 3 implies that each $\mathfrak{Z}_m$ is an invariant subspace for the operators $\mathcal{L}_\lambda$, $\mathcal{Q}$ and $\mathcal{C}$ in (2.30). (The subspaces $\mathfrak{Z}_m$ can also be defined by means of (3.2) and the representation (3.4) on $\mathfrak{Z}$.) It follows that equation (2.30) can be studied on any of the Hilbert spaces $\mathfrak{Z}_m$ merely by the use of restrictions rather than projections. The subspaces $\mathfrak{Z}_m$ are the natural subspaces of $\mathfrak{Z}$ in which to work because "sheets" of solutions do not seem to arise in $\mathfrak{Z}_m$ and solutions of (2.30) in $\mathfrak{Z}_m$ have special symmetry properties.

Note also that in carrying out the Lyapunov-Schmidt method in Section 2 the pair of Hilbert spaces $\mathcal{H}$ and $\mathfrak{Z}$ can be replaced by any pair of invariant subspaces $\mathcal{H}_m$ and $\mathfrak{Z}_m$. Thus, it follows that the solutions w^* and z^* constructed in Theorem 2 may be considered to have corresponding symmetry properties because z^* belongs to $\mathfrak{Z}_m$ whenever w^* belongs to $\mathcal{H}_m$. By making use of the solutions of equation (*) obtained in [3] for various $\mathcal{H}_m$ subspaces one can construct in this way a variety of solutions of (2.30) having special symmetry properties. Such a construction together with Theorem 1 then yields a variety of classical solutions of the John shell equations having special symmetry properties. In particular, the above results show that each of the solutions constructed in [3] for (w,λ) near $(0,\lambda_1)$ corresponds to a classical solution of Problem S with the same symmetry and stability properties.

BIBLIOGRAPHY

1. S.S. Antman, "Ordinary differential equations of non-linear elasticity I: Foundations of the theories of non-linearly elastic rods and shells," Arch. Rational Mech. Anal. 61(1976), 307-351.

2. F. John, "Estimates for the derivatives of the stresses in a thin shell and interior shell equations," Comm. Pure Appl. Math. 18(1965), 235-267.

3. G.H. Knightly and D. Sather, "Buckled states of a spherical shell under uniform external pressure", Arch. Rational Mech. Anal. 72(1980), 315-380.

4. G.H. Knightly and D. Sather, "Applications of group representations to the buckling of spherical shells", Applications of Nonlinear Analysis in the Physical Sciences, edited by H. Amann et al., Pitman, London, 1981.

5. W.T. Koiter, "Elastic stability and post-buckling behavior," Nonlinear Problems, edited by R. Langer, University of Wisconsin Press, Madison, 1963.

6. W.T. Koiter, "On the nonlinear theory of thin elastic shells," Proc. Kon. Nederl. Akad. Wet. Amsterdam B69 (1966), 1-54.

7. W.T. Koiter, "The nonlinear buckling problem of a complete spherical shell under uniform external pressure," Proc. Kon. Nederl. Akad. Wet. Amsterdam B72 (1969), 40-123.

8. D.H. Sattinger, "Transformation groups and bifurcation at multiple eigenvalues", Bull. Amer. Math. Soc. 79(1973), 709-711.

9. D.H. Sattinger, "Group representation theory, bifurcation theory, and pattern formation," J. Funct. Anal. 28(1978), 58-101.

10. D.H. Sattinger, "Bifurcation from rotationally invariant states," J. Math. Phys. 19(1978), 1720-1732.

11. N.J. Vilenkin, Special Functions and the Theory of Group Representations, Translations of Mathematical Monographs, Vol.22, American Mathematical Society, Providence, 1968.

Contemporary Mathematics
Volume 4
1981

REPEATED EIGENVALUES IN MECHANICAL OPTIMIZATION PROBLEMS*

Kyung K. Choi
and
Edward J. Haug

ABSTRACT. Optimal design problems with constraints on natural frequency and buckling load (eigenvalues) are shown to have repeated eigenvalues at an optimum. Both finite dimensional systems governed by matrix equations and infinite dimensional systems governed by boundary-value problems are shown to yield repeated eigenvalues. It is shown that repeated eigenvalues have at most directional differentials. Recent directional derivative results for repeated eigenvalues and nondifferentiable optimization theory are then employed to develop valid optimality criteria. These criteria are then applied to finite and infinite dimensional optimal design problems with constraints on natural frequency and buckling load.

1. INTRODUCTION

It has recently been shown [1] that even in the optimization of single structural elements such as columns, repeated eigenvalues can arise and must be accounted for in necessary conditions of optimality. The landmark paper of Ref. 1 shows clearly that if the presence of the repeated eigenvalue is ignored or not realized, then fundamentally incorrect results may arise. The dangers of optimizing structures under buckling constraints had earlier been noted in Ref. 2, with the consequence of imperfection sensitivity of repeated eigenvalues clearly demonstrated.

1980 Mathematics Subject Classification. 73H05.

*Research supported by NSF Grant CMS 80-05677.

It is thus becoming clear that repeated eigenvalues may indeed occur at an optimum design [1] and that the consequence can be severe [2].

In this paper, mechanical system optimization problems are analyzed. It is shown that repeated eigenvalues for buckling load and natural frequency arise in constrained optimal design problems. A two degree of freedom spring-mass vibration example shows clearly that repeated eigenvalues arise for even systems whose governing equations are matrix equations. The distributed parameter buckling example of Ref. 1 is analyzed to demonstrate the methods used for systems whose behavior is described by boundary-value problems.

As very recently shown [3], optimization of structures with repeated eigenvalues can lead to singular solutions. In this paper, a theory of differentiation of eigenvalues of operator equations with respect to design variables [4,5] is used to show that a repeated eigenvalue is not generally differentiable with respect to design. An extension of results presented in Ref. 5 is used to derive valid formulas for directional derivatives and prove quasi-differentiability of a repeated eigenvalue.

An abstract necessary condition of optimality [6] is derived, using the differentiation theory of Refs. 4 and 5. The necessary condition is illustrated using the elementary two degree of freedom example. The necessary condition is then used to study necessary conditions for the distributed parameter buckling problem of Ref. 1.

2. EXAMPLES OF STRUCTURAL OPTIMIZATION PROBLEMS WITH REPEATED EIGENVALUES

2.1 Simple Spring-Mass Optimal Design Problem with Repeated Natural Frequencies

Consider the spring-mass system shown in Fig. 2.1. The eigenvalue problem for small amplitude vibration of the rigid body is

$$Ky \equiv \begin{bmatrix} 4k_1 + k_2 & k_2 \\ k_2 & 4k_1 + k_2 \end{bmatrix} y = \zeta \begin{bmatrix} 2 & 1 \\ 1 & 2 \end{bmatrix} y \equiv \zeta My \qquad (2.1)$$

where $\zeta = \frac{2\omega^2 m}{3}$, m is the mass of the bar, $I^2 = mL^2/12$ is its polar moment of inertia, and horizontal motion of the bar is ignored.

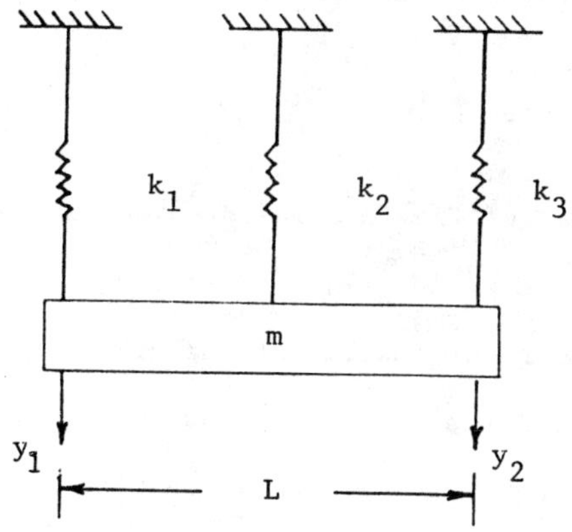

Figure 2.1 Two Degree of Freedom Spring-Mass System

The optimal design objective is to find design variables k_1 and k_2 to minimize weight of the spring supports, which is presumed to be of the form

$$\psi_0 = c_1k_1 + c_2k_2 \tag{2.2}$$

where c_1 and c_2 are known constants, subject to constraints that the eigenvalues ζ_1 and ζ_2 are no lower than $\zeta_0 > 0$ and the spring constants are nonnegative. In inequality constraint form, this is

$$\left.\begin{aligned}
\psi_1 &= \zeta_0 - \zeta_1 = \zeta_0 - \frac{4k_1 + 2k_2}{3} \leq 0 \\
\psi_2 &= \zeta_0 - \zeta_2 = \zeta_0 - 4k_1 \leq 0 \\
\psi_3 &= -k_1 \leq 0 \\
\psi_4 &= -k_2 \leq 0
\end{aligned}\right\} \tag{2.3}$$

Note that Eqs. 2.2 and 2.3 form a linear programming problem. The feasible set is shown graphically in Fig. 2.2. Note also that the slope of the line connecting points A and B in Fig. 2.2 is -2. The level lines of the cost function of Eq. 2.2 are straight, with slope equal to $-c_1/c_2$. The cost function decreases as level lines of the cost function move to the lower left. Thus, it is clear that point A (repeated eigenvalue) is the optimum design if $c_1/c_2 \geq 2$ and point B (simple eigenvalue) is the optimum design if $c_1/c_2 \leq 2$.

2.2 A Clamped-Clamped Distributed Column with Repeated Eigenvalues

The clamped-clamped column of Fig. 2.3 was studied as an optimal design problem by Tadjbakhsh and Keller [7]. They sought a cross-sectional area u(x), where moment of inertia of the cross section is $I = \alpha u^2$, to maximize the fundamental buckling load for a given volume of material in the column. Thus, the problem is to find $u \in L^\infty(0,1)$ that minimizes

$$\psi_0(u) = -\zeta(u) \tag{2.4}$$

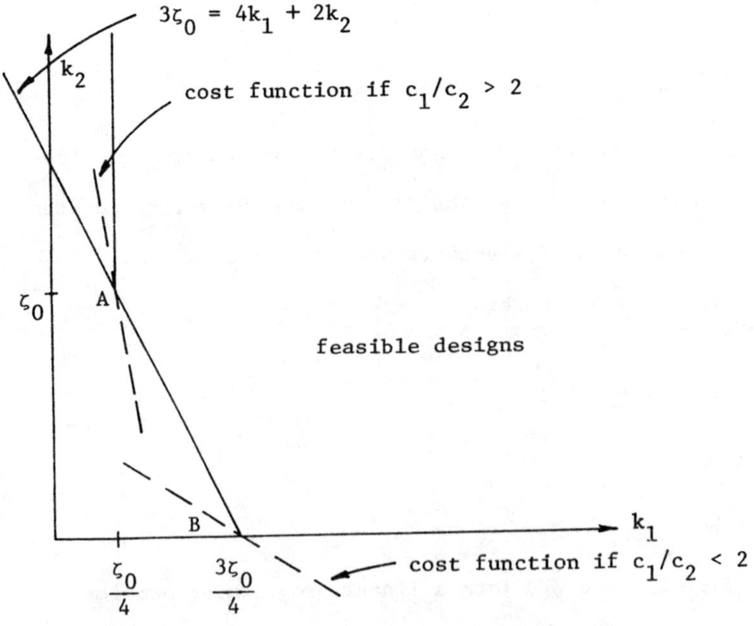

Figure 2.2 Feasible Region in Design Space for 2 DOF System

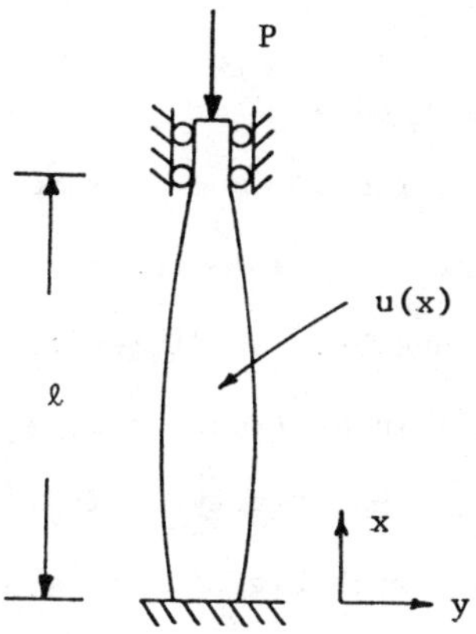

Figure 2.3 Column

subject to the volume constraint

$$\psi_1(u) = \int_0^\ell u\,dx - V = 0 \tag{2.5}$$

The buckling load ζ is the smallest eigenvalue of

$$\left.\begin{aligned} Ay &\equiv (E\alpha u^2(x)y'')'' = -\zeta y'' \equiv \zeta By \\ y(0) &= y'(0) = y(\ell) = y'(\ell) = 0 \end{aligned}\right\} \tag{2.6}$$

where $' \equiv \frac{d}{dx}$, E and α are material and geometrical constants, and

$$\phi = c - u(x) \le 0 , \qquad \text{a.e. in } [0,\ell] \tag{2.7}$$

It has recently been shown by Olhoff and Rasmussen [1] that one should expect that the fundamental mode is bimodal at the optimum design, for some values of $c > 0$ in Eq. 2.7. This prototype structural optimization problem is shown in Refs. 4 and 5 to be typical of structural optimization problems involving plates and planar elasticity, where the operators A and B are elliptic differential operators. Since $u \in L^\infty(0,\ell)$, it is clear that distributional solutions of the operator equation of Eq. 2.6 are required. In variational notation, this is

$$a_u(y,v) \equiv \int_0^\ell E\alpha u^2 y''v''dx = \zeta \int_0^\ell y'v'dx \equiv \zeta b_u(y,v) \tag{2.8}$$

with $y,v \in H_0^2(0,\ell)$.

3. NECESSARY CONDITIONS OF OPTIMALITY

3.1 Directional Derivatives of Eigenvalues

For broad classes of structures, including the examples of Section 2, it is shown in Ref. 4 that the operators are strongly elliptic, the bilinear forms a_u and b_u of Eq. 2.8 are Fréchet differentiable with respect to u, and their differentials in a direction h, $a^{(1)}_{u,h}$ and $b^{(1)}_{u,h}$, are relatively bounded by a_u and b_u, respectively. It is shown in Ref. 5 that if the eigenvalue ζ is simple, with eigenfunction y normalized by $b_u(y,y) = 1$, then ζ is Fréchet differentiable with respect to u, and its Fréchet differential in direction h is

$$\zeta^{(1)}_{u,h} = a^{(1)}_{u,h}(y,y) - \zeta b^{(1)}_{u,h}(y,y) \tag{3.1}$$

which, in the case of the column of Section 2.2, is

$$\zeta^{(1)}_{u,h} = \int_0^\ell 2E\alpha uh(y'')^2\, dx \tag{3.2}$$

If, on the other hand, $\zeta = \zeta_1 = \dots = \zeta_m$ is an m-fold eigenvalue, with eigenfunctions $y^i(x)$ orthonormalized so that $b_u(y^i,y^j) = \delta_{ij}$, then it is shown in Ref. 5 that the ζ_i, as functions of u, have at most directional differentials $\zeta^{(1)}_{u,h,i}$, which are eigenvalues of the matrix

$$M(u,h) = \left[a^{(1)}_{u,h}(y^i,y^j) - \zeta b^{(1)}_{u,h}(y^i,y^j) \right] \tag{3.3}$$

Since the eigenvalues of M depend on h, they define the bifurcation of the repeated eigenvalue ζ into possible simple eigenvalues as u is changed to u + h, for h small.

In the case of the column with a pair of repeated eigenvalues and b_u-orthonormal eigenfunctions y^1 and y^2,

$$M = 2E\alpha \begin{bmatrix} \int_0^\ell uh(y^{1''})^2\, dx & \int_0^\ell uhy^{1''}y^{2''}\, dx \\ \int_0^\ell uhy^{1''}y^{2''}\, dx & \int_0^\ell uh(y^{2''})^2\, dx \end{bmatrix} \tag{3.4}$$

3.2 Necessary Conditions of Optimality

In the case of a simple eigenvalue, the cost functional of Eq. 2.4 is differentiable and one may define a convex cone K of feasible directions, for example in the case of the column, for the constraint of Eq. 2.7 as

$$K = \{h \in L^{\infty}: \quad h = \gamma(u - u_0) \text{ for } u \text{ such that } c - u \leq 0 \text{ and } \gamma > 0\} \tag{3.5}$$

Now, an optimality condition for u_0 to be optimum is that [6, p. 83]

$$\lambda_0 \psi^{(1)}_{u_0,h,0} + \lambda_1 \psi^{(1)}_{u_0,h,1} \geq 0 \tag{3.6}$$

for some λ_0 and λ_1, not both zero, and all $h \in K$. In the case of the column, this is

$$\int_0^{\ell} [-\lambda_0 2E\alpha u(y'')^2 + \lambda_1] \, hdx \geq 0 \tag{3.7}$$

The situation is more complicated in case of a repeated eigenvalue. The directional derivative of $\zeta(u)$ is the smallest eigenvalue of the matrix of Eq. 3.3. But this is just

$$\zeta^{(1)}_{u_0,h} = \min_{\substack{\xi \in R^m \\ \xi^T\xi=1}} \xi^T \mathcal{M}(u_0,h)\, \xi \tag{3.8}$$

For fixed ξ, $\xi^T \mathcal{M} \xi$ is linear in h, since $\mathcal{M}$ is linear in h, so $\min_{\xi} \xi^T \mathcal{M} \xi$ is concave. Thus, the directional derivative $\psi^{(1)}_{u_0,h,0}$ of Eq. 2.4 is convex and an optimality condition for the repeated eigenvalue problem is, from Eq. 3.6,

$$-\lambda_0 \min_{\substack{\xi \in R^m \\ \xi^T\xi=1}} \sum_{i,j=1}^{m} \left[\xi_i \xi_j \left(a^{(1)}_{u_0,h}(y^i,y^j) - \zeta b^{(1)}_{u_0,h}(y^i,y^j)\right)\right] + \lambda_1 \psi^{(1)}_{u_0,h,1} \geq 0 \tag{3.9}$$

for all $h \in K$.

For the column with a double eigenvalue, this is

$$-\lambda_0 \, 2E\alpha \min_{\substack{\xi \in R^2 \\ \xi^T\xi=1}} \int_0^{\ell} u_0(\xi_1 y^{1''} + \xi_2 y^{2''})^2 \, hdx + \lambda_1 \int_0^{\ell} hdx \geq 0 \tag{3.10}$$

for all $h \in K$, where y^1 and y^2 is any b_u-orthonormal pair of eigenfunctions of Eq. 2.8 associated with $\zeta(u_0)$. Use of this condition for determination of an optimum design u appears to be nontrivial.

3.3 Optimality Conditions for the Spring-Mass Example

It is of interest to verify the optimality criterion of Eq. 3.6 for the simple spring-mass example of Section 2.1, at the solution with a repeated eigenvalue; $k_1 = \zeta_0/4$, $k_2 = \zeta_0$. The minimum in the right-hand side of Eq. 3.8 is the smaller of the eigenvalues of the matrix M, which are $4h_1/3 + 2h_2/3$ and $4h_1$, so Eq. 3.6 is, using the differential of ψ_0 from Eq. 2.2 and $\lambda_0 = 1$ (see Section 4.5),

$$c_1h_1 + c_2h_2 - \lambda_1 \min\{4h_1/3 + 2h_2/3,\ 4h_1\} \ge 0 \tag{3.11}$$

Consider first the case $4h_1 = 4h_1/3 + 2h_2/3$; i.e., $h_2 = 4h_1$. Then Eq. 3.11 is

$$c_1h_1 + 4c_2h_1 \ge \lambda_1 4h_1$$

for all $h_1 \in R^1$. Thus, equality must hold and

$$\lambda_1 = c_1/4 + c_2 \ge 0 \tag{3.12}$$

Consider second the case $4h_1 > 4h_1/3 + 2h_2/3$; i.e., $h_2 < 4h_1$. Then Eq. 3.11 is

$$c_1h_1 + c_2h_2 - \lambda_1(4h_1/3 + 2h_2/3) \ge 0$$

Using λ_1 from Eq. 3.12, this may be written

$$(c_1 - 2c_1)h_2 \le (c_1 - 2c_2)4h_1$$

Since this must hold for all h with $h_2 < 4h_1$, it is necessary that $c_1 - 2c_2 \ge 0$, or

$$c_1/c_2 \ge 2 \tag{3.13}$$

Finally, consider the case $4h_1 < 4h_1/3 + 2h_2/3$; i.e., $h_2 > 4h_1$. Then Eq. 3.11, with λ_1 from Eq. 3.12, becomes

$$c_2h_2 \ge c_2(4h_1)$$

Since this must hold for all h with $h_2 > 4h_1$, it is necessary that

$$c_2 \geq 0 \tag{3.14}$$

The necessary conditions thus reduce to the requirements of Eqs. 3.13 and 3.14, which simply require that $c_i \geq 0$, $i = 1,2$, and Eq. 3.13, which is precisely the condition determined in Section 2.1. Thus, the necessary condition of Eq. 3.6 is sharp for this problem; i.e., it is both necessary and sufficient (see Section 4.4) for an optimum design.

The foregoing optimality conditions, specifically Eq. 3.6, may seem to be unnecessarily complicated. This viewpoint is reinforced by the well-known optimization results obtained by Prager, Taylor, and Niordson, using variational formulationgs with Lagrange multipliers in problems with a simple eigenvalue [8,9]. However, formal Lagrange multiplier method which uses the Rayleigh quotient characterization of the eigenvalues does not give valid necessary conditions [10].

4. DERIVATION OF NECESSARY CONDITIONS OF OPTIMALITY FOR DISCRETE SYSTEMS

4.1 Introduction

In Section 3, necessary conditions of optimality were derived, using the fact that m-fold eigenvalues have directional derivatives [5]. However, use of those necessary conditions for determination of an optimum design appears to be difficult (e.g., Eq. 3.10). In this section, it is shown that the smallest eigenvalue of buckling and vibration problems is quasi-differentiable [6, p. 68]. The quasi-differentiability of the smallest eigenvalue enables one to derive more useful necessary conditions.

In this section, the structure to be optimized is first approximated by finite element methods. The eigenvalue problem that arises in vibration or buckling of structures is then of the form

$$A(u)y = \zeta\, B(u)y \tag{4.1}$$

where $u \in R^n$, $y \in R^k$, and A(u) and B(u) are k × k matrices. In the problems considered here, A(u) and B(u) are symmetric and positive definite.

The statement of the problem is as follows: find $u \in R^n$ that minimizes the cost function $\psi_0(u)$, subject to Eq. 4.1 and the constraints

$$\psi_1(u) = \zeta_0 - \zeta(u) \leq 0 \tag{4.2}$$

$$u \in Q = \{u \in R^n : c_i - u_i \leq 0,\ c_i > 0,\ i = 1,2,\ldots,n\} \tag{4.3}$$

Here it is assumed that the cost function $\psi_0(u)$ is differentiable and $\zeta(u)$ in Eq. 4.2 is the smallest eigenvalue of Eq. 4.1. It is assumed that the smallest eigenvalue has multiplicity $m \geq 1$ at the optimum point $u_0 \in Q$.

4.2 Quasi-Differentiable Functionals

It is shown in Ref. 6 (p. 42) that if $\psi(u)$ is a real convex functional on a Banach space U, then the directional differential

$$\psi^{(1)}_{u_0,h} = \lim_{t \to +0} \frac{\psi(u_0 + th) - \psi(u_0)}{t} \tag{4.4}$$

exists for all u_0 and h, and is given by

$$\psi^{(1)}_{u_0,h} = \sup_{m \in \partial\psi(u_0)} m(h) \tag{4.5}$$

where $\partial\psi(u_0)$ is the subdifferential of ψ at u_0, which is defined as

$$\partial\psi(u_0) = \{m \in U^* : \psi(u) - \psi(u_0) \geq m(u - u_0) \text{ for all } u \in U\} \tag{4.6}$$

Moreover, $\partial\psi(u_0)$ is non-empty, convex, weak* closed, and bounded in U^* [6, p. 39].

A real functional $\psi(u)$ on a Banach space U is said to be quasi-differentiable at a point u_0 if there exists a convex, weak* closed, and bounded set $\partial\psi(u_0) \subset U^*$ such that Eq. 4.5 holds. Note that if $\psi(u)$ is quasi-differentiable, then its directional derivative $\psi^{(1)}_{u_0,h}$ is convex in h, since it is the supremum of linear functionals. The class of quasi-differentiable functionals contains all Fréchet and Gateaux differentiable functionals. Also, Eq. 4.5 shows that all convex functionals are quasi-differentiable.

If $\psi_1(u)$ in Eq. 4.2 is convex in u, then it is quasi-differentiable at $u_0 \in Q$. Hence, it is natural to ask whether $\psi_1(u)$ of Eq. 4.2 is convex or not. Of course the convexity of $\psi_1(u)$ depends on the operators A(u) and B(u) of each given problem. Since A(u) and B(u) of the problems considered here are positive definite, one can find the smallest eigenvalue as

$$\zeta(u) = \inf_{\substack{y \in D_A \\ y \neq 0}} \frac{(A(u)y,y)}{(B(u)y,y)} \tag{4.7}$$

Since the supremum of convex functionals is convex [11, p. 168]; if, for fixed $y \in D_A$, $-(A(u)y,y)/(B(u)y,y)$ is convex in u, then $\psi_1(u)$ is convex. For example, in the spring-mass example of Section 2.1, $-(A(k)y,y)/(B(k)y,y)$ is linear in k, so $\psi_1(k)$ in Eq. 4.2 is convex. However, $\psi_1(u)$ is not always convex in u. A counter example is the clamped-clamped column problem of Section 2.2, where

$$\zeta(u) = \inf_{\substack{y \in D_A \\ y \neq 0}} \frac{\int_0^{\ell} E\alpha u^2 y''^2 dx}{\int_0^{\ell} y'^2 dx} \tag{4.8}$$

Let v = 2u and let $\gamma_1 = \gamma_2 = 1/2$. Then, $\psi_1(\gamma_1 u + \gamma_2 v) \geq \gamma_1 \psi_1(u) + \gamma_2 \psi_1(v)$ and ψ_1 is not convex.

4.3 Quasi-Differentiability of the Eigenvalue

In this section, it is to be shown that $\psi_1(u)$ in Eq. 4.2 is quasi-differentiable at $u_0 \in Q$. To do so, first note that $\psi_1(u)$ is directionally differentiable at u_0 and $\psi^{(1)}_{u_0,h,1} = -\zeta^{(1)}_{u_0,h}$, where $\zeta^{(1)}_{u_0,h}$ is the smallest eigenvalue of the matrix [5]

$$M(u_0,h) = \left[\left(A^{(1)}_{u_0,h}\, y^i,\, y^j \right) - \zeta(u_0) \left(B^{(1)}_{u_0,h}\, y^i,\, y^j \right) \right] \tag{4.9}$$

and $\{y^i\}_{i=1,2,\dots,m}$ is any B(u)-orthonormal basis of the eigenspace associated with $\zeta(u_0)$. Using the Rayleigh quotient, one has

$$\psi^{(1)}_{u_0,h,1} = \sup_{\substack{\xi \in R^m \\ \|\xi\|=1}} (-M(u_0,h)\xi,\xi) \tag{4.10}$$

which is convex in h, since it is the supremum of linear functionals.

To show that $\psi_1(u)$ is quasi-differentiable at u_0, one has to find a convex, closed, and bounded set $\partial\psi_1(u_0) \subset R^n$ such that

$$\psi^{(1)}_{u_0,h,1} = \sup_{m \in \partial\psi_1(u_0)} (m,h) \tag{4.11}$$

Let $N = \{\xi \in R^m : \|\xi\| = 1\}$. As a functional of $\xi \in N$ and $h \in R^n$, $(-M(u_0,h)\xi,\xi)$ in Eq. 4.10 has the Fréchet differential with respect to h, $(-M(u_0,e_i)\xi,\xi)_{i=1,2,\dots,n}$, and satisfies the hypotheses of theorem 3.5 of Ref. 6, where N is compact and the e_i are vectors in R^n with all zero elements except the ith, which is one. Thus $\psi^{(1)}_{u_0,h,1}$, as a functional of h, is quasi-differentiable at h = 0. Moreover, the subdifferential of $\psi^{(1)}_{u_0,h,1}$ at h = 0 is

$$\partial\psi^{(1)}_{u_0,h,1}(0) = \{m \in R^n : (m,h) = \int_N (-M(u_0,h)\xi,\xi)\,d\mu\,(\xi)$$

$$= \int_N \sum_{i=1}^n (-M(u_0,e_i)\xi,\xi)\,h_i\,d\mu\,(\xi)\} \tag{4.12}$$

where $d\mu\,(\xi)$ denotes a nonnegative measure whose total variation on N is one; $\int_N d\mu\,(\xi) = 1$. Also, $\partial\psi^{(1)}_{u_0,h,1}(0)$ is convex, closed, and bounded (see proof of theorem 3.5 of Ref. 6) in R^n.

Now,

$$(m,h) = \int_N (-M(u_0,h)\xi,\xi)\,d\mu\,(\xi) \le \sup_{\xi \in N} (-M(u_0,h)\xi,\xi) \tag{4.13}$$

On the other hand, by taking $d\mu\,(\xi)$ to be a Dirac measure at $\xi \in N$,

$$\psi^{(1)}_{u_0,h,1} = \sup_{\xi \in N} (-M(u_0,h)\xi,\xi) = \sup_{m \in \partial\psi^{(1)}_{u_0,h,1}(0)} (m,h) \tag{4.14}$$

Hence, $\psi_1(u)$ is quasi-differentiable at u_0, with $\partial\psi_1(u_0) = \partial\psi^{(1)}_{u_0,h,1}(0)$. Note that for a general non-convex functional $\psi(u)$, with a directional derivative, the subdifferential of the functional at u_0 can be defined to be $\partial\psi^{(1)}_{u_0,h}(0)$ [11, p. 196].

4.4 Necessary Conditions of Optimality

Necessary conditions of optimality are now derived by evaluating the derivatives of the cost functional and constraint functionals. In order to apply theorem 4.1 of Ref. 6, one has to construct a convex cone K_Q, such that for any $h \in K_Q$, $u_\lambda = u_0 + \lambda h$ [6, p. 88] belongs to Q, for sufficiently small $\lambda > 0$. Indeed, define

$$K_Q = \{h \in R^n : h = \gamma(u - u_0),\ u \in Q,\ \gamma > 0\} \tag{4.15}$$

Then $u_\lambda = u_0 + \lambda\gamma(u - u_0) = (1 - \lambda\gamma)\, u_0 + \lambda\gamma u \in Q$ for $\lambda \in [0, 1/\gamma]$, because Q is a convex set. Now by theorem 4.1 of Ref. 6, the necessary condition is that there exist $\lambda_0 \geq 0$ and $\lambda_1 \geq 0$, not both zero, such that

$$\lambda_0 \sum_{i=1}^{n} \frac{\partial\psi_0}{\partial u_i}(u_0)\, h_i - \lambda_1 \zeta^{(1)}_{u_0,h}$$

$$= \lambda_0 \sum_{i=1}^{n} \frac{\partial\psi_0}{\partial u_i}(u_0)\, h_i - \lambda_1 \inf_{\xi \in N} (M(u_0,h)\xi,\xi) \geq 0 \tag{4.16}$$

for all $h \in K_Q$ and $\lambda_1(\zeta_0 - \zeta(u_0)) = 0$, which is the complementary slackness condition. It should be noted that the above expression is not linear in h.

Since $\psi_1(u)$ is quasi-differentiable, Eq. 4.14 holds and the necessary condition of Eq. 4.16 is equivalent to

$$\lambda_0 \sum_{i=1}^{n} \frac{\partial\psi_0}{\partial u_i}(u_0)\, h_i + \lambda_1 \sup_{m \in \partial\psi^{(1)}_{u_0,h,1}(0)} (m,h) \geq 0 \tag{4.17}$$

for all $h \in K_Q$. Because $\partial\psi^{(1)}_{u_0,h,1}(0)$ is bounded, by lemma 4.2 of Ref. 6 the necessary condition of Eq. 4.17 is equivalent to the existence of an $m \in \partial\psi^{(1)}_{u_0,h,1}(0)$ (that is, existence of a measure $d\mu(\xi)$) such that

$$\lambda_0 \sum_{i=1}^{n} \frac{\partial\psi_0}{\partial u_i}(u_0)\, h_i + \lambda_1 \sum_{i=1}^{n} \int_N (M(u_0,e_i)\xi,\xi)\,d\mu(\xi)\;\; h_i \geq 0 \tag{4.18}$$

for all $h \in K_Q$. That is, there exists an $f \in R^n$ with components

$$f_i = \lambda_0 \frac{\partial\psi_0}{\partial u_i}(u_0) - \lambda_1 \int_N (M(u_0,e_i)\xi,\xi)\; d\mu\,(\xi) \tag{4.19}$$

that is an element of the dual cone K_Q^* of K_Q, where

$$K_Q^* = \{f \in R^n : (f,h) \geq 0 \text{ for all } h \in K_Q\} \tag{4.20}$$

and the necessary condition of Eq. 4.18 is equivalent to, by definition of K_Q,

$$(f,\, u - u_0) = \sum_{i=1}^{n} f_i \left(u_i - u_{0i}\right) \geq 0 \tag{4.21}$$

for all $u \in Q$.

It can be shown that the necessary condition of Eq. 4.21 is equivalent to [10]

$$\left.\begin{aligned} f_i &= \lambda_0 \frac{\partial\psi_0}{\partial u_i}(u_0) - \lambda_1 \int_N (M(u_0,e_i)\xi,\xi)\; d\mu\,(\xi) \geq 0, \text{ if } i \in I(u_0) \\ f_j &= \lambda_0 \frac{\partial\psi_0}{\partial u_j}(u_0) - \lambda_1 \int_N (M(u_0,e_j)\xi,\xi)\; d\mu\,(\xi) = 0, \text{ if } j \notin I(u_0) \end{aligned}\right\} \tag{4.22}$$

where $I(u_0) = \{i : 1 \leq i \leq n,\; u_{0i} = c_i\}$.

Now if the design variable constraint is everywhere active before the eigenvalue constraint is active, then by the complementary slackness condition, $\lambda_1 = 0$ and the system is optimized. Hence one can assume that the design variable constraint is not everywhere active. Thus, λ_1 will not be zero. If it were, λ_0 will be zero in Eq. 4.22, unless $\partial\psi_0/\partial u_i\,(u_0) \geq 0$ for all $i \in I(u_0)$ and $\partial\psi_0/\partial u_j\,(u_0) = 0$ for all $j \notin I(u_0)$. But if $\partial\psi_0/\partial u_j\,(u_0) = 0$ for all $j \notin I(u_0)$, then one can eliminate the u_j's from the design variable vector and, in this case, the design variable constraint is everywhere active. Hence, one assumes $\partial\psi_0/\partial u_j\ (u_0) \neq 0$ for all $j \notin I(u_0)$.

The next question is "when can one let $\lambda_0 = 1$?" Note in Eq. 4.16, if $\lambda_0 = 0$, then $-\lambda_1 \zeta^{(1)}_{u_0,h} \geq 0$ and $\zeta^{(1)}_{u_0,h} \leq 0$ for all $h \in K_Q$. Hence, if one shows that there exists an $h \in K_Q$ such that $\zeta^{(1)}_{u_0,h} > 0$, then one can set $\lambda_0 = 1$. Using $I(u_0)$, K_Q in Eq. 4.15 can be expressed as

$$K_Q = \{h \in R^n : h_i \geq 0 \text{ for } i \in I(u_0),\ h_j \in R \text{ for } j \notin I(u_0)\} \tag{4.23}$$

Define a subset of K_Q as

$$\overline{K} = \{h \in R^n : h_i = 0 \text{ for } i \in I(u_0),\ h_j \in R \text{ for } j \notin I(u_0)\} \tag{4.24}$$

Then $\overline{K}$ is a subspace of R^n. In case the design constraint is nowhere active, $I(u_0) = \emptyset$ and $K_Q = \overline{K} = R^n$. Consider two cases [10];

(A) If $M(u_0,e_j) = 0$ for any $j \notin I(u_0)$, then $\lambda_0 = 0$, by Eq. 4.22.

(B) If $M(u_0,e_j) \neq 0$ for any $j \notin I(u_0)$, then there is an $h \in \overline{K} \subset K_Q$ such that $\zeta^{(1)}_{u_0,h} \neq 0$, which in turn implies that there exists an $h \in \overline{K} \subset K_Q$ such that $\zeta^{(1)}_{u_0,h} > 0$. Hence one can let $\lambda_0 = 1$.

From (A) and (B), if $M(u_0,e_j) = 0$ for any $j \notin I(u_0)$, then $M(u_0,e_j) = 0$ for all $j \notin I(u_0)$ and if $M(u_0,e_j) \neq 0$ for any $j \notin I(u_0)$, then $M(u_0,e_j) \neq 0$ for all $j \notin I(u_0)$. Moreover, if case (B) holds, then there exists an $h \in K_Q$ such that $\zeta^{(1)}_{u_0,h} > 0$. Hence there exists a $u \in Q$ such that $\psi_1(u) < 0$, which is the Slater condition.

In Section 4.2 it was shown that the convexity of $\psi_1(u)$ depends on the operators $A(u)$ and $B(u)$ of the given problem. Suppose case (B) holds, then if $\psi_1(u)$ of a given problem is convex, the necessary condition of Eq. 4.22 is also a sufficient condition of optimality [11, p. 69], whereas if $\psi_1(u)$ is not convex, it may not be a sufficient condition.

However, even if $\psi_1(u)$ is not convex, one can show [10] that up to first order approximation, it is impossible to have an improved design that satisfies the necessary condition.

4.5 Spring-Mass Example

It is of interest to verify the necessary condition of optimality of Eq. 4.22 for the simple spring-mass example of Section 2.1 at the optimum point, $k_0 = \left[\frac{\zeta_0}{4}, \zeta_0\right]^T$. Using M-orthonormal eigenvectors $y^1 = \frac{1}{\sqrt{2}}[1,0]^T$ and $y^2 = \frac{1}{\sqrt{6}}[1,-2]^T$, one has

$$M(k_0,e_1) = \begin{bmatrix} 2 & \frac{2}{\sqrt{3}} \\ \frac{2}{\sqrt{3}} & \frac{10}{3} \end{bmatrix} \qquad M(k_0,e_1) = \begin{bmatrix} \frac{1}{2} & -\frac{1}{2\sqrt{3}} \\ -\frac{1}{2\sqrt{3}} & \frac{1}{6} \end{bmatrix} \tag{4.25}$$

Hence, $M(k_0,e_1) \neq 0$ and $M(k_0,e_2) \neq 0$, so one can let $\lambda_0 = 1$, which justifies Eq. 3.11.

Now the necessary condition of Eq. 4.22 is

$$\left.\begin{aligned} c_1 - \lambda_1 \int_N \xi^T M(k_0,e_1)\xi \, d\mu\,(\xi) &= 0 \\ c_2 - \lambda_1 \int_N \xi^T M(k_0,e_2)\xi \, d\mu\,(\xi) &= 0 \end{aligned}\right\} \tag{4.26}$$

Since $\int_N d\mu\,(\xi) = 1$, from above equation,

$$\begin{aligned} c_1 - 2c_2 &= \lambda_1 \int_N \xi^T \begin{bmatrix} 1 & \sqrt{3} \\ \sqrt{3} & 3 \end{bmatrix} \xi \, d\mu\,(\xi) \\ &\geq \lambda_1 \inf_{\xi \in N} \xi^T \begin{bmatrix} 1 & \sqrt{3} \\ \sqrt{3} & 3 \end{bmatrix} \xi \geq 0 \end{aligned} \tag{4.27}$$

Hence $c_1 - 2c_2 \geq 0$ and $c_i \geq 0$, $i = 1,2$, which is precisely the condition determined in Section 2.1.

5. DERIVATION OF NECESSARY CONDITIONS OF OPTIMALITY FOR DISTRIBUTED SYSTEMS

5.1 Introduction

In this section, methods of Section 4 are extended to distributed parameter systems, for which design and response variables of the system are functions. As in Section 4, it is to be shown that the smallest eigenvalue of a structural system is quasi-differentiable. Using this quasi-differentiability, pointwise necessary conditions are derived.

In this section, the state equations are given in variational form, since the design variable u is in $L^{\infty}(\Omega)$ [5]. The general form of variational state equation is

$$a_u(y,v) = \zeta b_u(y,v), \text{ for all } v \in V(\Omega) \tag{5.1}$$

where the physical domain Ω is a bounded open subset of R^n, $V(\Omega)$ is a Hilbert space that is dense in $L^2(\Omega)$, $a_u(y,v)$ is a strongly elliptic bilinear form on $V(\Omega)$, and $b_u(y,v)$ is a bilinear form on $L^2(\Omega)$.

The statement of the problem is as follows; find $u \in L^{\infty}(\Omega)$ that minimizes the cost function

$$\psi_0(u) = \int_{\Omega} F(u(x))\, dx \tag{5.2}$$

subject to Eq. 5.1 and

$$\psi_1(u) = \zeta_0 - \zeta(u) \le 0 \tag{5.3}$$

$$u \in Q = \{u(x) \in L^{\infty}(\Omega) : c - u(x) \le 0 \text{ a.e. in } \Omega,\ c > 0\} \tag{5.4}$$

Here it is assumed that $\zeta(u)$ is the smallest eigenvalue of Eq. 5.1 and that the smallest eigenvalue has multiplicity $m \ge 1$ at the optimum design $u_0 \in Q$. It is assumed that $F(u)$ in Eq. 5.2 is continuous in u, differentiable with respect to u, and $F_u(u)$ is continuous in u. Then, $\psi_0(u)$ is Fréchet differentiable [10] and

$$\psi^{(1)}_{u_0,h,0} = \int_{\Omega} F_u(u_0(x))\, h(x)\, dx \tag{5.5}$$

5.2 Quasi-Differentiability of the Eigenvalue

In this section, it is to be shown that $\psi_1(u)$ in Eq. 5.3 is quasi-differentiable at $u_0 \in Q$. To show this, first note that $\psi_1(u)$ is directionally differentiable at u_0 and $\psi^{(1)}_{u_0,h,1} = -\zeta^{(1)}_{u_0,h}$, where $\zeta^{(1)}_{u_0,h}$ is the smallest eigenvalue of the matrix [5]

$$M(u_0,h) = \left[a^{(1)}_{u_0,h}(y^i,y^j) - \zeta(u_0)\, b^{(1)}_{u_0,h}(y^i,y^j)\right] \tag{5.6}$$

and $\{y^i\}_{i=1,2,\ldots,m}$ is any b_u-orthonormal basis of the eigenspace of $\zeta(u_0)$. Using the Rayleigh quotient,

$$\psi^{(1)}_{u_0,h,1} = \sup_{\substack{\xi \in R^m \\ ||\xi||=1}} (-M(u_0,h)\xi,\xi) \tag{5.7}$$

which is convex in h, being the supremum of linear functionals.

To prove that $\psi_1(u)$ is quasi-differentiable at u_0, one has to find a convex, weak* closed, and bounded set $\partial\psi_1(u_0) \subset (L^\infty(\Omega))^*$ such that

$$\psi^{(1)}_{u_0,h,1} = \sup_{m \in \partial\psi_1(u_0)} (m,h) \tag{5.8}$$

Let $N = \{\xi \in R^m : ||\xi|| = 1\}$. Then as a functional of $\xi \in N$ and $h \in L^\infty(\Omega)$, $(-M(u_0,h)\xi,\xi)$ in Eq. 5.7 satisfies the hypothesis of theorem 3.5 of Ref. 6, with Fréchet differential with respect to h equal to $(-M(u_0,\cdot)\xi,\xi)$, where N is compact. Thus $\psi^{(1)}_{u_0,h,1}$, as a functional of h, is quasi-differentiable at h = 0. Moreover, the subdifferential of $\psi^{(1)}_{u_0,h,1}$ at h = 0 is

$$\partial\psi^{(1)}_{u_0,h,1}(0) = \{m \in (L^\infty(\Omega))^* : (m,h) = \int_N (-M(u_0,h)\xi,\xi)\, d\mu\,(\xi)\} \tag{5.9}$$

where $d\mu\,(\xi)$ denotes a nonnegative measure whose total variation on N is equal to one; i.e., $\int_N d\mu\,(\xi) = 1$. Also, $\partial\psi^{(1)}_{u_0,h,1}(0)$ is convex, weak* closed, and bounded (see proof of theorem 3.5 of Ref. 6).

Now,

$$(m,h) = \int_N (-M(u_0,h)\xi,\xi)\, d\mu\,(\xi) \le \sup_{\xi \in N} (-M(u_0,h)\xi,\xi) \tag{5.10}$$

On the other hand, by taking $d\mu(\xi)$ to be a Dirac measure at $\xi \in N$,

$$\psi^{(1)}_{u_0,h,1} = \sup_{\xi \in N} (-M(u_0,h)\xi,\xi) = \sup_{m \in \partial\psi^{(1)}_{u_0,h,1}(0)} (m,h) \tag{5.11}$$

Hence $\psi_1(u)$ is quasi-differentiable at u_0, with $\partial\psi_1(u_0) = \partial\psi^{(1)}_{u_0,h,1}(0)$. In fact, as in the discrete case, $\partial\psi^{(1)}_{u_0,h,1}(0)$ can be defined to be the subdifferential of the functional $\psi_1(u)$ at u_0 [11].

5.3 Necessary Conditions of Optimality

As in the discrete case, in order to apply theorem 4.1 of Ref. 6, one has first to construct a convex cone K_Q such that for any $h \in K_Q$, $u_\lambda = u_0 + \lambda h$ [6, p. 88] belongs to Q, for all $\lambda > 0$ sufficiently small. As in the discrete case, one can set

$$K_Q = \{h \in L^\infty(\Omega) : h = \gamma(u(x) - u_0(x)),\ u \in Q,\ \gamma > 0\} \tag{5.12}$$

This is true, since $u_\lambda = u_0 + \lambda\gamma(u - u_0) = (1 - \lambda\gamma)\,u_0 + \lambda\gamma u \in Q$ for $\lambda \in [0,1/\gamma]$ and Q is a convex set. By theorem 4.1 of Ref. 6, the necessary condition of optimality is that there exist $\lambda_0 \geq 0$ and $\lambda_1 \geq 0$, not both zero, such that

$$\lambda_0 \int_\Omega F_u(u_0(x))\, h(x)\, dx - \lambda_1\, \zeta^{(1)}_{u_0,h}$$

$$= \lambda_0 \int_\Omega F_u(u_0(x))\, h(x)\, dx - \lambda_1 \inf_{\xi \in N} (M(u_0,h)\xi,\xi) \geq 0 \tag{5.13}$$

for all $h \in K_Q$. Moreover, $\lambda_1(\zeta_0 - \zeta(u_0)) = 0$, which is the complimentary slackness condition. Note that the above expression is not linear in h.

By Eq. 5.11, the necessary condition of Eq. 5.13 is equivalent to

$$\lambda_0 \int_\Omega F_u(u_0(x))h(x)dx + \lambda_1 \sup_{m \in \partial\psi^{(1)}_{u_0,h,1}(0)} (m,h) \geq 0 \tag{5.14}$$

for all $h \in K_Q$. Since $\partial\psi^{(1)}_{u_0,h,1}$ is bounded, by lemma 4.2 of Ref. 6, the above condition is equivalent to the existence of an $m \in \partial\psi^{(1)}_{u_0,h,1}(0)$ (i.e., existence of a measure $d\mu\,(\xi)$) such that $f \in (L^\infty(\Omega))^*$, defined by

$$\begin{aligned}(f,\bar{u}) &= \lambda_0 \int_\Omega F_u(u_0(x))\,\bar{u}(x)\,dx + \lambda_1\,(m,\bar{u}) \\ &= \lambda_0 \int_\Omega F_u(u_0(x))\,\bar{u}(x)\,dx - \lambda_1 \int_N (M(u_0,\bar{u})\xi,\xi)\,d\mu\,(\xi)\end{aligned} \tag{5.15}$$

for all $\bar{u} \in L^\infty(\Omega)$, is an element of dual cone K_Q^*. That is,

$$(f,h) = \lambda_0 \int_\Omega F_u(u_0(x))\,h(x)\,dx - \lambda_1 \int_N (M(u_0,h)\xi,\xi)\,d\mu\,(\xi) \geq 0 \tag{5.16}$$

for all $h \in K_Q$. By definition of K_Q, the necessary condition of Eq. 5.16 is equivalent to

$$\begin{aligned}(f,u-u_0) = \lambda_0 &\int_\Omega F_u(u_0(x))(u(x)-u_0(x))\,dx \\ &- \lambda_1 \int_N (M(u_0,u-u_0)\xi,\xi)\,d\mu\,(\xi) \geq 0\end{aligned} \tag{5.17}$$

for all $u \in Q$. Let $\bar{M}_{ij}(u_0,\bar{u})$ denote the integrands of the elements of the matrix $M(u_0,\bar{u})$; i.e.,

$$M_{ij}(u_0,\bar{u}) = \int_\Omega \bar{M}_{ij}(u_0,\bar{u})\,dx \tag{5.18}$$

Then Eq. 5.15 can be expressed as

$$\begin{aligned}(f,\bar{u}) &= \lambda_0 \int_\Omega F_u(u_0(x))\,\bar{u}(x)\,dx - \lambda_1 \int_N \int_\Omega (\bar{M}(u_0,\bar{u})\xi,\xi)\,dx\,d\mu\,(\xi) \\ &= \lambda_0 \int_\Omega F_u(u_0(x))\,\bar{u}(x)\,dx - \lambda_1 \int_\Omega \bar{u}(x) \int_N (\bar{M}(u_0,1)\xi,\xi)\,d\mu\,(\xi)\,dx \\ &= \int_\Omega \left[\lambda_0 F_u(u_0(x)) - \lambda_1 \int_N (\bar{M}(u_0,1)\xi,\xi)\,d\mu\,(\xi)\right] \bar{u}(x)\,dx\end{aligned} \tag{5.19}$$

where Fubini's theorem [12, p. 150] is used in the second equality. Thus,

$$f(x) = \lambda_0 F_u(u_0(x)) - \lambda_1 \int_N (\bar{M}(u_0,1)\xi,\xi)\,d\mu\,(\xi) \tag{5.20}$$

and $f \in L^1(\Omega)$. By the same method used in example 10.5 of Ref. 13 (in Ref. 13, $\Omega \subset R$ was considered which is extended to $\Omega \subset R^n$ in Ref. 10), the necessary condition of Eq. 5.17 is equivalent to

$$\left(\lambda_0 F_u(u_0(x)) - \lambda_1 \int_N (\overline{M}(u_0,1)\xi,\xi)\, d\mu\,(\xi)\right)\ [u(x) - u_0(x)] \geq 0 \tag{5.21}$$

for all $u \in Q$ and almost all $x \in \Omega$.

It can be shown that the necessary condition of Eq. 5.21 is equivalent to [10]

$$f(x) = \lambda_0\, F_u(u_0(x)) - \lambda_1 \int_N (\overline{M}(u_0,1)\xi,\xi)\, d\mu\,(\xi) \begin{cases} \geq 0 \text{ a.e. in } \Omega\backslash I(u_0) \\ = 0 \text{ a.e. in } I(u_0) \end{cases} \tag{5.22}$$

where $I(u_0) = \{x \in \Omega : u_0(x) > c \text{ a.e.}\}$, which is the set where the design variable constraint is not active, almost everywhere.

Now if the design variable constraint is active almost everywhere, before the eigenvalue constraint is active, then by the complementary slackness condition $\lambda_1 = 0$ and the system is optimized. Hence, one can assume that the design variable constraint is not almost everywhere active. As in the discrete case, assume $F_u(u_0(x)) \neq 0$ a.e. in $I(u_0)$. Then λ_1 will not be zero, because otherwise, λ_0 will be zero in Eq. 5.22, which is a contradiction.

As in the discrete case, it is to be determined when one can let $\lambda_0 = 1$. In Eq. 5.13, if $\lambda_0 = 0$, then $-\lambda_1 \zeta^{(1)}_{u_0,h} \geq 0$ and $\zeta^{(1)}_{u_0,h} \leq 0$ for all $h \in K_Q$. Hence, if one shows that there exists an $h \in K_Q$ such that $\zeta^{(1)}_{u_0,h} > 0$, then one can let $\lambda_0 = 1$. The set K_Q in Eq. 5.12 can be written as

$$K_Q = \{h \in L^\infty(\Omega) : h(x) \geq 0 \text{ a.e. in } \Omega\backslash I(u_0) \text{ and } h(x) \in R \text{ a.e. in } I(u_0)\} \tag{5.23}$$

Define a subset of K_Q as

$$\overline{K} = \{h \in L^\infty(\Omega) : h(x) = 0 \text{ a.e. in } \Omega\backslash I(u_0) \text{ and } h(x) \in R \text{ a.e. in } I(u_0)\} \tag{5.24}$$

which is a subspace of $L^\infty(\Omega)$. Note that if the design variable constraint is not active almost everywhere, then $I(u_0) = \Omega$ and $K_Q = \overline{K} = L^\infty(\Omega)$. Consider two cases [10];

(A) If $\overline{M}(u_0,1) = 0$ for any subset of $I(u_0)$ with strictly positive measure, then $\lambda_0 = 0$ by Eq. 5.22.

(B) If $\overline{M}(u_0,1) \neq 0$ for any subset of $I(u_0)$ with strictly positive measure, then there is an $h \in \overline{K} \subset K_Q$ such that $\zeta^{(1)}_{u_0,h} \neq 0$, which in turn implies that there exists an $h \in \overline{K} \subset K_Q$ such that $\zeta^{(1)}_{u_0,h} > 0$. Hence one can let $\lambda_0 = 1$.

From (A) and (B), either $\overline{M}(u_0,1) = 0$ a.e. in $I(u_0)$ or $\overline{M}(u_0,1) \neq 0$ a.e. in $I(u_0)$. Moreover, if case (B) holds, then there exists an $h \in K_Q$ such that $\zeta^{(1)}_{u_0,h} > 0$. Hence there exists a $u \in Q$ such that $\psi_1(u) < 0$, which is the Slater condition.

When case (B) holds, if $\psi_1(u)$ of Eq. 5.3 is convex, then the necessary condition of Eq. 5.22 is also a sufficient condition of optimality [11, p. 69], whereas if $\psi_1(u)$ is not convex, it may not be sufficient.

Even if $\psi_1(u)$ is not convex, if case (B) holds, then one can prove [10] that it is not possible, up to the first order approximation, to have a better design than u_0, which satisfies the necessary condition.

5.4 Clamped-Clamped Column Example

In this section, the buckling problem of the clamped-clamped column of Section 2.2 [1,7] is analyzed, using the necessary conditions developed in the previous sections. Instead of looking for $u \in L^\infty(0,\ell)$, as in Section 2.2, u will be restricted to be in $C(0,\ell)$, as in Refs. 1 and 7. Introducing the dimensionless variables

$$\overline{u}(x) = \frac{u(x)\ell}{V}, \quad \overline{\zeta} = \frac{3\ell^4}{E\alpha V^2} \tag{5.25}$$

where the coordinate x is the non-dimensionalized by ℓ, the column problem of Section 2.2 becomes; find $u \in C(0,1)$ that minimizes

$$\psi_0(u) = -\zeta(u) \tag{5.26}$$

subject to the volume constraint

$$\psi_1(u) = \int_0^1 u(x)\, dx - 1 = 0 \tag{5.27}$$

and design variable constraint

$$Q = \{u \in C(0,1) : c \le u(x),\ c > 0\} \tag{5.28}$$

where $\zeta(u)$ is the smallest eigenvalue of

$$\left.\begin{aligned} &(u^2(x)y'')'' = -\zeta y'' \\ &y(0) = y'(0) = y(1) = y'(1) = 0 \end{aligned}\right\} \tag{5.29}$$

In variational notation, Eq. 5.29 is

$$a_u(y,v) \equiv \int_0^1 u^2(x) y'' v'' dx = \zeta \int_0^1 y'v' dx = \zeta b_u(y,v) \tag{5.30}$$

It is assumed that at the optimum design $u_0(x)$, the eigenvalue $\zeta(u_0)$ is double. Note that Eq. 5.27 is an equality constraint and $\psi_1(u)$ is a linear functional. Thus, a corollary of Ref. 6 (p. 88) applies and one can set

$$K_Q = \{h \in C(0,1) : h = \gamma(u(x) - u_0(x)),\ u \in Q,\ \gamma > 0\} \tag{5.31}$$

and the same method used in Section 5.3 applies. Hence, the necessary condition is that there exist $\lambda_0 \ge 0$ and λ_1, not both zero, such that

$$-\lambda_0 \zeta^{(1)}_{u_0,h} + \lambda_1 \int_0^1 h(x)\, dx = -\lambda_0 \inf_{\xi \in N} (\mathcal{M}(u_0,h)\xi,\xi) + \lambda_1 \int_0^1 h(x)\, dx$$

$$= \lambda_0 \sup_{m \in \partial \psi^{(1)}_{u_0,h,0}(0)} (m,h) + \lambda_1 \int_0^1 h(x)\, dx \ge 0 \tag{5.32}$$

for all $h \in K_Q$, where $\partial \psi^{(1)}_{u_0,h,0}(0)$ is the subdifferential of the functional $\psi_0(u)$ at u_0. By proceeding in the same way as in Section 5.3, one gets the equivalent pointwise necessary conditions

$$-\lambda_0 \int_N (\overline{\mathcal{M}}(u_0,1)\xi,\xi)\, d\mu(\xi) + \lambda_1 \begin{cases} \ge 0 \text{ in } [0,1]\backslash I(u_0) \\ = 0 \text{ in } I(u_0) \end{cases} \tag{5.33}$$

where $I(u_0) = \{x \in [0,1] : u_0(x) > c\}$, $\xi_1^2 + \xi_2^2 = 1$, and $\overline{\mathcal{M}}(u_0,1)$ is the integrand of the matrix $M(u_0,1)$,

$$\overline{M}(u_0,1) = 2u_0(x) \begin{bmatrix} (y^{1''})^2 & y^{1''}y^{2''} \\ y^{1''}y^{2''} & (y^{2''})^2 \end{bmatrix} \tag{5.34}$$

If $0 < c < 1$, then the design variable constraint will not be active everywhere and λ_0 will not be zero. If it were zero, λ_1 will be zero by Eq. 5.33, which is a contradiction. Thus, one can set $\lambda_0 = 1$. Also, since $\overline{M}(u_0,1) \neq 0$ for any subset of $I(u_0)$, $\lambda_1 \neq 0$ (see Section 5.3). In fact, since $\overline{M}(u_0,1)$ is positive semi-definite for each x, $\lambda_1 > 0$ by Eq. 5.33.

Define a function $\nu(x)$ such that $\nu(x) \geq 0$ if $u_0(x) = c$ and $\nu(x) = 0$ if $u_0(x) > c$. Then the necessary condition of Eq. 5.33 becomes

$$\int_N (\overline{M}(u_0,1)\xi,\xi)\, d\mu\,(\xi) - \lambda_1 + \nu(x) = 0 \tag{5.35}$$

The necessary condition of Eq. 5.35 is different from the result of Olhoff and Rasmussen [1]. Note that

$$\int_N (M(u_0,h)\xi,\xi)\, d\mu\,(\xi) = \int_N \xi^T \begin{bmatrix} M_{11}(u_0,h) & 0 \\ 0 & {}_{22}(u_0,h) \end{bmatrix} \xi d\mu\,(\xi)$$

$$= M_{11}(u_0,h) \int_N \xi_1^2\, d\mu\,(\xi) + M_{22}(u_0,h) \int_N \xi_2^2\, d\mu\,(\xi) \tag{5.36}$$

for all $h \in K_Q$ if and only if $M_{12}(u_0,h) = M_{21}(u_0,h) = 0$, for all $h \in K_Q$. Let $\int_N \xi_1^2\, d\mu\,(\xi) = 1 - \gamma$, $0 \leq \gamma \leq 1$. Then $\int_N \xi_2^2\, d\mu\,(\xi) = \int_N (1 - \xi_1^2)\, d\mu\,(\xi) = \gamma$. Thus, the necessary condition of Eq. 5.32 is equivalent to

$$-\lambda_0 \left[(1 - \gamma)\, M_{11}(u_0,h) + \gamma M_{22}(u_0,h)\right] + \lambda_1 \int_0^1 h(x)\, dx \geq 0 \tag{5.37}$$

for all $h \in K_Q$, if and only if $M_{12}(u_0,h) = 0$ for all $h \in K_Q$. By the same procedure used in Section 5.3, the functional necessary condition of Eq. 5.37 is equivalent to the pointwise necessary condition (with $\lambda_0 = 1$)

$$2u_0(x) \left[(1-\gamma)(y^{1''})^2 + \gamma(y^{2''})^2\right] - \lambda_1 + \nu(x) = 0 \tag{5.38}$$

where $\nu(x) \geq 0$ if $u_0(x) = c$ and $\nu(x) = 0$ if $u_0(x) > c$. Equation 5.38 is the result of Ref. 1. Hence the necessary condition of Eq. 5.35 is equivalent to Eq. 5.38 if and only if

$$M_{12}(u_0,h) = \int_0^1 2u_0(x)\, y^{1''} y^{2''}\, h(x)\, dx = 0 \tag{5.39}$$

for all $h \in K_Q$.

Since the boundary conditions at both ends are identical and the design variable constraint is symmetric with respect to the center of the column, the area of the optimum design will be symmetric. Hence, one can restrict the variations $h \in K_Q$ to be symmetric. Also, at the optimum, one can choose two b_u-orthonormal eigenfunctions, which give the same eigenvalue, so that one is symmetric and the other is antisymmetric. Then $u_0(x)y^{1''}y^{2''}$ $h(x)$ is antisymmetric and Eq. 5.39 is satisfied, for all symmetric $h \in K_Q$. Hence in this case the result of Ref. 1 is obtained.

REFERENCES

1. Olhoff, N. and Rasmussen, S.H., "On Single and Bimodal Optimum Buckling Loads of Clamped Columns," Int. J. Solids and Structures, 13 (1977), pp. 605-614.

2. Thompson, J.M.T. and Hunt, C.W., "Dangers of Structural Optimization," Engineering Optimization, 2 (1974), pp. 99-110.

3. Masur, E.F. and Mroz, Z., "Singular Solutions in Structural Optimization Problems," Proceedings, IUTAM Symposium on Variational Methods in the Mechanics of Solids, (Ed. S. Nemat-Nasser), Pergamon Press, New York, 1980, 337-343.

4. Haug, E.J. and Rousselet, B., "Design Sensitivity Analysis in Structural Mechanics I: Static Response Variations," J. Str. Mech., 8, No. 1 (1980), pp. 17-41.

5. Haug, E.J. and Rousselet, B., "Design Sensitivity Analysis in Structural Mechanics II: Eigenvalue Variations," J. Str. Mech., 8, No. 2 (1980), pp. 161-186.

6. Pschenichnyi, B.N., Necessary Conditions for an Extremum, Marcel-Dekker, New York, 1971.

7. Tadjbakhsh, I. and Keller, J.B., "Strongest Columns and Isoparametric Inequalities for Eigenvalues," J. of Applied Mechanics, 29 (1962), pp. 159-164.

8. Prager, W. and Taylor, J.E., "Problems of Optimal Structural Design," J. Applied Mechanics, 35, No. 1, 1968, pp. 102-106.

9. Niordson, F.I., "On the Optimal Design of a Vibrating Beam," Quarterly Applied Math., 24, No. 1 (1965), pp. 47-53.

10. Choi, K.K. and Haug, E.J., "Optimization of Structures with Repeated Eigenvalues," Optimization of Distributed Parameter Structures (Eds. E.J. Haug and J. Cea), Sijthoff & Noordhoff, Alphen aan den Rijn, Netherlands, 1981.

11. Ioffe, A.D. and Tihomirov, V.M., Theory of Extremal Problems, North-Holland Publishing Co., New York, 1979.

12. Rudin, W., Real and Complex Analysis, McGraw-Hill, Inc., New York, 1974.

13. Girsanov, I.V., Lecture Notes in Economics and Mathematical Systems, Springer-Verlag, New York, 1972.

MATERIALS DIVISION
COLLEGE OF ENGINEERING
THE UNIVERSITY OF IOWA
IOWA CITY, IOWA 52242

Contemporary Mathematics
Volume 4
1981

EFFECTS OF SOME NONLINEAR TERMS ON THE BUCKLING OF ELASTIC BODIES*

Vadim Komkov and E. J. Haug

ABSTRACT. In a recent article [1] the authors have indicated the importance of considering the so-called "small" nonlinear terms to analyze the optimization of a buckled column. In this article this analysis is continued, detached from the optimization problem. Some obvious deficiencies of the linear theory are pointed out, and a comparison is made between the predictions of the two theories.

0. NOTATION

E - Young's modulus, which is considered to be a positive constant

A(x) - cross-sectional area

I(x) - moment of inertia of cross-sectional area about the neutral axis

ρ - radius of curvature

ν - Poisson's ratio (which is a constant $0 < \nu < 1/2$)

y(x) - linear deflection

θ(x) - angular deflection

1. SOME HISTORICAL REMARKS

The "usual" analysis of stability of compressed struts basically follows the arguments offered by Euler in his fundamental works on the subject. Euler's discovery of the buckling phenomenon, which appeared in his 1743 article, was followed by a more comprehensive analysis presented in

1980 Mathematics Subject Classification. 73H05.

*Research Supported by NSF Grant No. CMS 80-05677

1757 (See C. Truesdell [2]). The linear equation which is presented in this paper in the form

$$(EIy'')'' + Py'' = 0 \tag{1.1}$$

with appropriate boundary conditions appears in a slightly different form and in a different notation in the work of Euler. We shall follow Euler's arguments, but we shall simplify the notation and terminology and utilize the Kirchhoff energy approach, in what can be considered as a typical classroom derivation of 1.1.

2. THE LINEAR THEORY

We shall consider the simplest case of a compressed column, which is simply supported (by swivel joints) at both ends and is subjected to a concentrated load P applied at the ends, as shown on Figure 1. The shape may remain straight y(x) describes the shape of the compressed column) or it may change to some unknown shape $y = y(x)$. The boundary conditions for the problem illustrated in Fig. 1 are

$$y(0) = y(\overline{\ell}) = y''(0) = y''(\overline{\ell}) = 0 \tag{2.1}$$

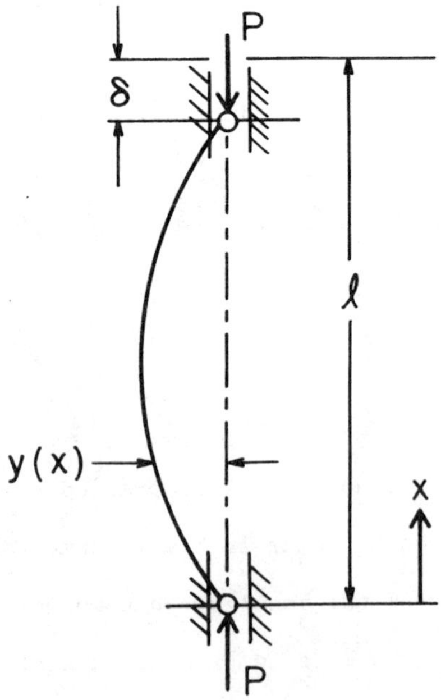

Figure 1. Hinged Column

Let us recall some of the assumptions of the linear theory. The bending moment caused by the force P at the point with coordinate x is given by

$$M(x) = - P\, y(x) \tag{2.2}$$

Using a variety of assumtions such as there exists a neutral surface, plane sections remain plane, and stresses vary linearly with the distance from the neutral surface, one can derive the basic equation of elastic beam theory, relating the radius of curvature of the neutral axis p to the magnitude of the bending moment:

$$\rho^{-1} = M(x)/EI(x) \tag{2.3}$$

An elementary geometric equality is given by

$$\rho^{-1} = y''/[1+(y')^2]^{3/2} \tag{2.4}$$

which is valid for any functions y(x) having continuously differentiable second derivatives.

The arguments supporting the linear theory rely on the "smallness" of y'. Hence, equating the denominator of the right hand-side expression in Eq. 2.4 to one and substituting Eq. 2.4 into Eq. 2.3, one obtains the approximate formula

$$EI(x)y''(x) = M(x) \tag{2.5}$$

The corresponding approximate formula for the strain energy of the bent beam is given by

$$U = \frac{1}{2} \int_0^{\ell-\delta} EI\,(x)\,(y'')^2\,dx \tag{2.6}$$

Usually, the axial deflection δ is ignored and the strain energy is written in the form

$$U = \frac{1}{2} \int_0^{\ell} EI(x)\,(y'')^2\,dx \tag{2.7}$$

This elementary review is given to stress the simplifying assumptions made in the derivation of Eqs. 1.1 to 2.7 and to stress the circumstances under which such assumptions can no longer be realistically made.

The assumptions listed above are standard assumptions made in the linear theory of elasticity for bent beams. Other assumptions that carry significant theoretical consequences are made with regard to the mechanism of buckling. Let us briefly repeat the classroom arguments, which distantly echo Euler's paper of 1757. (For a typical presentation paralleling our arguments, but in full detail, see for example Den Hartog [3]). Let ℓ denote the length of the column just prior to the loss of stability. If we assume that the length of the column does not change, due to the change from a straight shape $y(x) \equiv 0$ to the shape described by the function $y = y(x)$, the contraction in the direction of the force is described by

$$\delta_G = \int_0^\ell [(1 + (y')^2/2 + (y')^4/8 + \ldots) - 1]\, dx$$

$$= \int_0^\ell \left((y')^2/2 + (y')^4/8 + \ldots\right) dx \approx \frac{1}{2}\int_0^\ell (y')^2\, ds \tag{2.8}$$

where the approximation is only valid if terms of order four and higher are ignored. The effects of shear and the reduction in compressive strain energy, because the compressive force no longer acts in the axial direction, are ignored in this analysis.

Difficulties encountered in optimization of design for a column are discussed by the authors in [1]. Also see [4], [5], [6]. The existence of singularities in optimal designs has been partially explained in [5] and [7]. The linear theory that results from minimization of the quadratic function $P\delta_G - U = - V$, (where δ_G is given by Eq. 2.8 and U by Eq. 2.7 is reduced to the solution of the eigenvalue problem of Eq. 1.1 in the setting of the Sobolev space H_0^2. The difficulties encountered by J.B. Keller [8] and other early writers who considered the design optimization problem in this setting (see [9],[10], [11]) were explained by appearance of multiple eigenvalues. The problem of minimizing the Raleigh quotient

$$\lambda_1 = P_{crit} = \min_{y \in H_0^2[0,\ell]} \left\{ \int_0^\ell EI(x)(y'')^2 dx \Big/ \int_0^\ell (y')^2 dx \right\} \qquad (2.9)$$

fails to yield the correct solution when the "obvious" technique of differentiating λ_1 with respect to y(x) is applied, setting the value of the first derivative equal to zero. The reason for this failure is the appearance at some point in the minimization procedure of multiple eigenvalues. The function λ_1 fails to be Fréchet differentiable with respect to the vector y(x) in H_0^2 and formal differentiation procedure leads to predictable difficulties.

Simple examples are offered by K.K. Choi and E.J. Haug [12] illustrating this serious theoretical difficulty. These examples, which offer a discussion of the simplest two-degrees of freedom system consisting of a mass and two springs, illustrate fully the problems that arise in a formal approach to the design optimization of more complex elastic systems.

In the paper [1], the authors have pointed out that inclusion of some so called "small" nonlinear terms can completely alter the theoretical analysis and prevent the appearance of singularities in optimal design of slender columns. This idea is more extensively investigated in the present paper.

The basic statement, which is mathematically correct for the problem posed by the eigenvalue equation of Eq. 1.1 with appropriate boundary conditions such as those in Eq. 2.1; i.e., either y = 0, or there exist a discrete set of eigenvalues P_i and corresponding functions $y_i(x)$ for which this problem has a solution, does not correspond to the physical reality. Given any P>0, which is not an eigenvalue of Eqs. 1.1 and 2.1, there exists a physical solution of the corresponding loading problem within the framework of elastic behavior of the column; i.e., the correct mathematical model must have a continuous spectrum of acceptable values of P, within some range $\overline{P} \geq P \geq 0$, where $\overline{P}$ is assigned to prevent the necessity of studying the more complex inelastic behavior of the buckled column.

3. TWO SIMPLE EXAMPLES

The authors believe that any theory should first be tested on the simplest possible cases, to gain an insight into its advantages, difficulties and shortcomings. The simplest example, modelling the behavior of a buckled strut, which offers a meaningful interpretation of the difficulties arising in distributed parameter systems, is illustrated in Fig. 2.

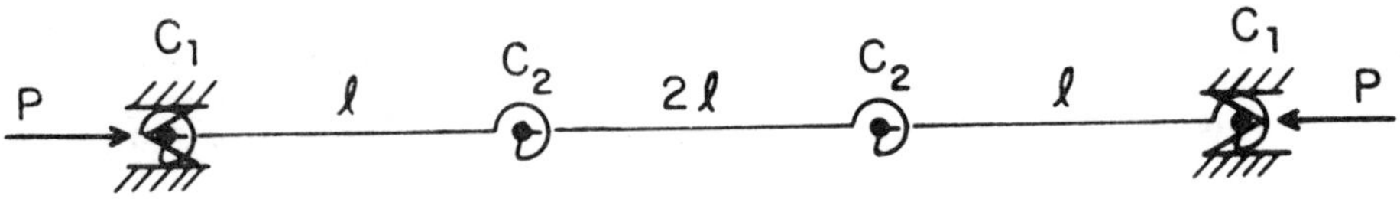

Figure 2. A Bar Model of The Hinged Column

The model consists of three rigid bars, connected by swivel pin joints containing torsional springs. The spring constants are C_1 and C_2 at the end and internal joints, respectively. This is the simplest, and crudest finite element model of the column problem illustrated on Fig. 1. Elastic models of beams represented by rigid bars that are connected by pin joints with torsional springs are well-known to numerical analysts working in structural design. The simplicity of the model shown in Fig. 2 permits us to illustrate some difficulties of the linear model, while performing fairly elementary computations. Even here, the complexity of non-linear behavior leads to difficult numerical schemes. Unfortunately, further simplification of our model was not possible. For example, the model shown on Fig. 3 does not permit computation of multiple eigenvalues. It therefore does not offer a meaningful contrast between the linear and non-linear behavior in this respect.

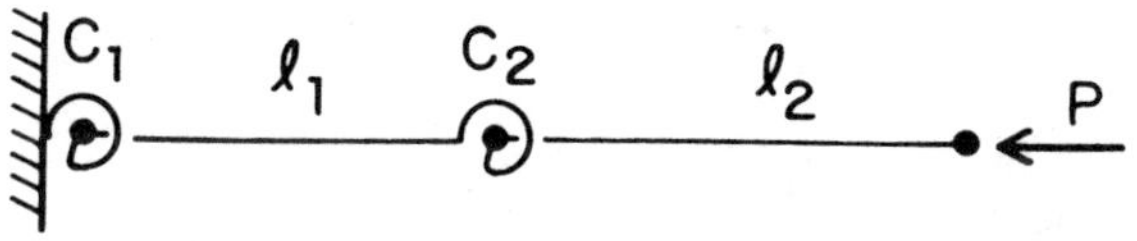

Figure 3. A Bar Model of the Cantilevered Column

We refer to Fig. 2 and compute the buckled configurations. If symmetry is assumed, the shapes are shown on Figs. 4 and 5, respectively. In the configuration shown in Fig. 4, the strain energy stored in the torsional springs at the joints is given by

$$V = (C_1 + C_2)\,\alpha_1^2 \tag{3.1}$$

The work done by the force P_1 is

$$W = P_1(\ell - \ell\cos\alpha_1) \approx P_1\ell\alpha_1^2/2 \tag{3.2}$$

if we linearize, where the approximation is obtained by assuming that the work is a quadratic functional of α_1.

If V = W, then

$$P_1 = 2(C_1 + C_2)/\ell \tag{3.3}$$

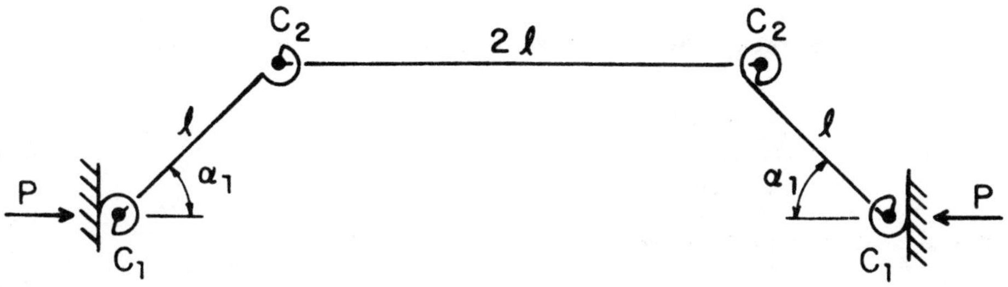

Figure 4. Symmetric Mode

In the configuration shown in Fig. 5, the strain energy is given by

$$V = C_1 \alpha_2^2 + 4C_2\alpha_2^2 \tag{3.4}$$

The work performed by the force P_2 is given by

$$W = 2P_2 (\ell-\ell\cos \alpha_2) \approx P_2 \ell\alpha_2^2 \tag{3.5}$$

Equating strain energy and the work done, we obtain

$$P_2 = (C_1 + 4C_2)/\ell \tag{3.6}$$

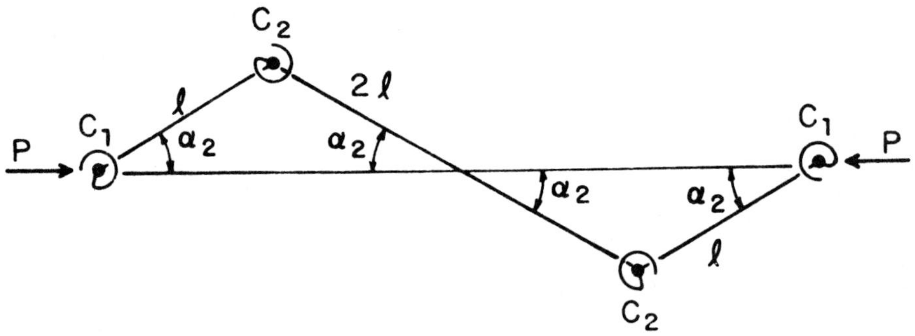

Figure 5. Asymmetric Mode

The repeated eigenvalue problem occurs in the linear theory when we choose spring constants C_1 and C_2 such that

$$C_1 = 2C_2 .$$

The linear theory predicts nothing about the actual behavior of the buckled structure. Either shape (Fig. 4 or 5) or any linear combination of these shapes is acceptable.

In the nonlinear theory the configurations shown on Fig. 4 occurs when

$$P_1\ell = \frac{(C_1 + C_2)\alpha_1^2}{1 - \cos \alpha_1} \tag{3.7}$$

and the configuration shown on Fig. 5 occurs when

$$P_2\ell = \frac{(C_1 + 4C_2)\alpha_2^2}{1 - \cos \alpha_2} \tag{3.8}$$

Equating P_1 and P_2 , we obtain the relationship

$$2(C_1 + C_2)\alpha_1^2/(1-\cos\alpha_1) = (C_1 + 4C_2)\alpha_2^2/(1-\cos\alpha_2) \qquad (3.9)$$

At the design values of C_1 , C_2 such that $C_1 = 2C_2$, we obtain

$$\alpha_1^2/(1-\cos\alpha_1) = \alpha_2^2/(1-\cos\alpha_2) \qquad (3.10)$$

which implies (within a certain range of values), that $\alpha_1 = \alpha_2$. However, it is easy to check that both of these values do not satisfy equality of the work done and potential energy gained and, therefore (in general), do not describe within the quasistatic theory framework any equilibrium positions of this structure in the configurations shown by Figs. 4 and 5.

To predict an equilibrium path, following the onset of instability, we investigate the set of values $\{\alpha\}$ and the corresponding values of $P > \min\{P_1, P_2\}$, such that

$$\frac{\partial U}{\partial \alpha} = 0 \qquad (3.11)$$

where $U = V - W$. Suppose that $\alpha_1 > \alpha_2$.

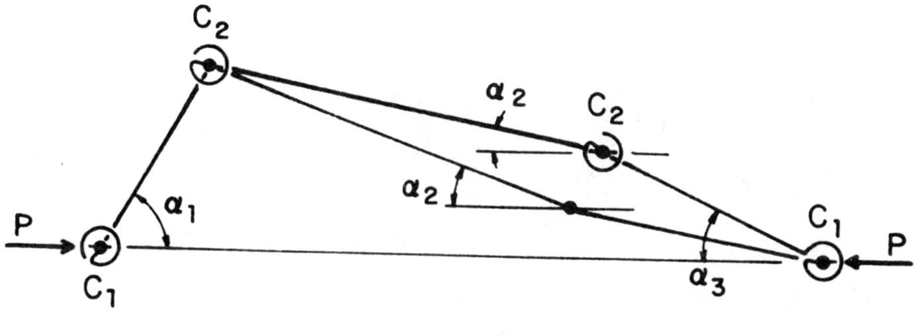

Figure 6. General Deflected Shape

Holding the total deflection (hence the work done is constant) it is easy to show that we actually decrease the potential energy by reducing α_1 and increasing α_3. This shows that the configuration shown in Fig. 6 did not correspond to a stationary value of potential energy. It shows, in fact, that the minimum potential energy configuration has to be symmetric. Hence, Figs. 4 and 5 were drawn correctly.

Modelling the Compressibility Effects

We retain the scheme shown in Fig. 2, but include telescoping springs to show the effects of compressibility. This is illustrated in Fig. 7, where ℓ corresponds to the free length with the telescoping spring not compressed, and K denotes the usual linear spring constant.

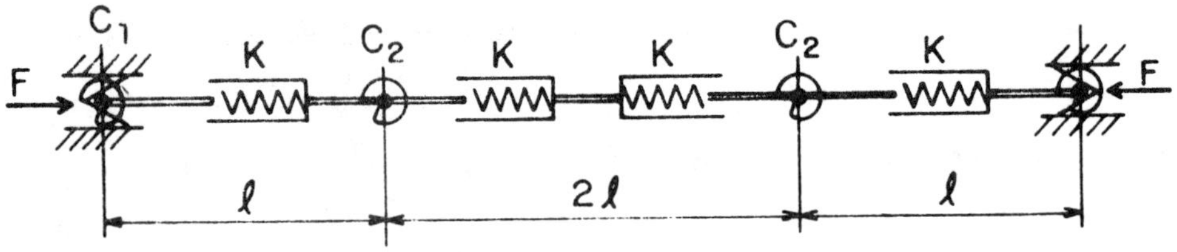

Figure 7.

The three bars are no longer rigid. An axial force F will produce deflection $K^{-1}F$ in the bars of length ℓ and $2K^{-1}F$ in the bar of length 2ℓ. If this linkage assumes a configuration shown on Fig. 4, then the deflection caused in the sloping bars is given by

$$\delta_S = 2K^{-1}P/\cos\alpha \tag{3.12}$$

The corresponding "shortening" in the horizontal direction is given by

$$\delta_1 = 2\ell - 2(\ell - K^{-1}P/\cos\alpha)\cos\alpha \tag{3.13}$$

The corresponding horizontal bar deflection is simply $2K^{-1}P = \delta_2$. The force P_1 thus displaces the structure by the amount

$$\delta_1 + \delta_2 = 2(\ell - \ell\cos\alpha_1) + 4K^{-1}P_1 \tag{3.14}$$

in the horizontal direction. The work done is

$$W = P_1\ell\ (1-\cos\alpha_1) + 2K^{-1}P_1{}^2 \tag{3.15}$$

The strain energy is given by

$$V = K^{-1}P_1^2\ /\cos^2\alpha_1 + 2K^{-1}P_1^2 + C_1\alpha_1^2 + C_2\alpha_1^2 \tag{3.16}$$

Equating V and W we obtain the equilibrium equation

$$(K^{-1}/\cos^2\alpha_1)\ P_1{}^2 = P_1\ell(1-\cos\alpha_1)+C_1\alpha_1{}^2+C_2\alpha_1{}^2 = 0 \qquad (3.17)$$

We solve for P_1 in terms of α_1, to obtain

$$P_1 = \frac{K\cos^2\alpha_1}{2}\left\{ \ell(1-\cos\alpha_1) \mp \left[\ell^2(1-\cos\alpha_1)^2 - 4K^{-1}(C_1\alpha_1^2 + C_2\alpha_1{}^2)/\cos^2\alpha_1\right]^{1/2}\right\} \qquad (3.18)$$

We shall denote by $P_{1,1}$ the smaller of the two values and by $P_{1,2}$ the larger one, if the quantity under the square root is positive. No equilibrium position exists if this quantity is negative; i.e., if

$$\frac{\ell^2K}{4(C_1+C_2)} < \frac{\alpha_1^2}{\cos^2\alpha_1\ (1-\cos\alpha_1)} \qquad (3.19)$$

If we expand the right hand side of Eq. 3.19 in infinite series, which converges uniformly and absolutely for all $\alpha < \pi/2$, we have

$$\frac{\ell^2K}{4(C_1+C_2)} < (1/2 - 13/12\ \alpha^2 + 29/72\ \alpha^4 - \cdots)^{-1} \qquad (3.20)$$

as a condition for nonexistence of an equilibrium position. If K is very small, no equilibrium point can exist in the vicinity of the trivial equilibrium point $P_1 = 0$, $\alpha_1 = 0$.

The largest value of α, $< \alpha < \frac{\pi}{2}$ for which

$$\frac{\ell^2K}{4(C_1+C_2)} = \frac{\alpha_1^2}{\cos^2\alpha_1(1-\cos\alpha_1)} \qquad (3.21)$$

will be called the threshold value and will be denoted by α^c . For any $\alpha > \alpha^c$, equilibrium does not exist. As subsequent equilibrium study conforms, the column suffers a catastrophic collapse.

The treshhold force corresponding to α^c is given by

$$P^c = 1/2\ K\ell\cos^2\alpha_1{}^c\ (1 - \cos\alpha_1^c\ ) \qquad (3.22)$$

We comment that the function under the square root sign in Eq. 3.18,

$$f(\ell, K^{-1}, C_1, C_2, \alpha_1) = \ell^2(1-\cos\alpha_1)^2 - 4K^{-1}(C_1\alpha_1^2 + C_2\alpha_1^2)/\cos^2\alpha_1 \qquad (3.23)$$

is not a monotone function of α_1 and that for some choices of K^{-1}, ℓ, and $(C_1 + C_2)$, its behavior may be as depicted in Fig. 8.

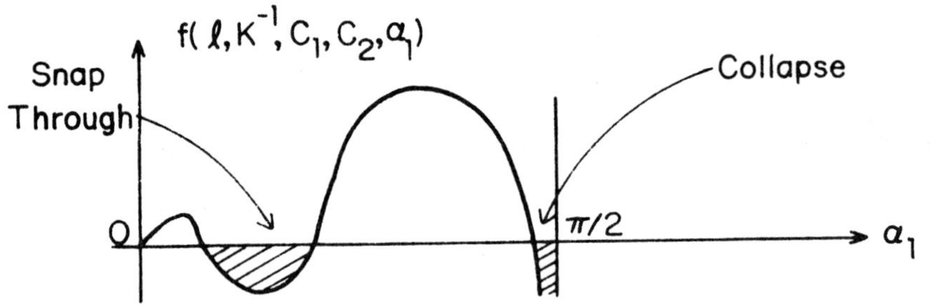

Figure 8. Variation of f

We should comment on our use of quasi-static assumptions, which tacitly imply that the work done is exactly equal to the strain energy of the deformed structure. We already indicated that this is a somewhat unrealistic hypothesis if sudden changes in the shape of a structure takes place, such as a "snap through" from one region in which stability is possible to another region of stability. Experimental studies of thin columns indicate the existence of such behavior. The fact that the quasi-static approach is unsatisfactory is confirmed by a simple calculation, comparing the values obtained from the equality V=W with the values obtained by local stationary behavior of the total potential energy. The differences in energy values must be credited to the dissipation of kinetic energy, which we could not account for in our theoretical model.

The stability of any configuration shown in Fig. 4 can be examined by looking at the derivative of the potential energy.

$$\frac{\partial U}{\partial \alpha_1} = -2K^{-1}P_1^2 \sin\alpha_1 \cos^{-3}\alpha_1 + 2(C_1 + C_2)\alpha_1 - 2P_1 \ell \sin\alpha_1 \qquad (3.24)$$

$$\frac{\partial U}{\partial P_1} = 2K^{-1}P_1/\cos^2\alpha_1 + 4K^{-1}P_1 - 2\ell(1-\cos\alpha_1)$$
$$-4K^{-1}P_1 \tag{3.25}$$

The Hessian matrix for U is

$$\begin{pmatrix} U_{\alpha_1\alpha_1} & U_{\alpha_1 P_1} \\ U_{P_1\alpha_1} & U_{P_1P_1} \end{pmatrix} \tag{3.26}$$

where

$$\left.\begin{aligned} U_{\alpha_1\alpha_1} &= -2K^{-1}P_1^2\cos^{-2}\alpha_1 + 6K^{-1}P_1^2\sin^2\alpha_1\cos^{-4}\alpha_1 \\ U_{\alpha_1P_1} &= -4K^{-1}P_1\sin\alpha_1\cos^{-3}\alpha_1 - 2\ell\sin\alpha_1 \\ U_{P_1P_1} &= +2K^{-1}\cos^{-2}\alpha_1 \end{aligned}\right\} \tag{3.27}$$

The condition

$$U_{\alpha_1\alpha_1}U_{P_1P_1} - U^2_{P_1\alpha_1} > 0 \tag{3.28}$$

implies a stable equilibrium at a point of stationary potential energy.

We compare the behavior of the system shown on Fig. 7, in a configuration illustrated in Fig. 5. Here, the force in the end member is $P/\cos\alpha_2$. The axial displacement is given by

$$\delta = 2P_2\ell\,(1-\cos\alpha_2) + 4K^{-1}P \tag{3.29}$$

The work done is

$$P_2\,\ell(1-\cos\alpha_2) + 2K^{-1}P_2^2 \tag{3.30}$$

The strain energy is

$$V = C_1\alpha_2^2 + 4C_2\alpha_2^2 + 2K^{-1}P_2^2 /\cos^2\alpha_2 \tag{3.31}$$

The quadratic equation in P_2 is now given by

$$2P_2^2 (\cos^{-2}\alpha_2 - 1) - K P_2 \ell(1-\cos\alpha_2) - K(C_1^1 \alpha_2^2 + 4C_2\alpha_2^2) = 0$$

Thus,

$$P_2 = (\tfrac{1}{4}\cot^2\alpha_2)\left\{K\ell (1-\mathrm{Cos}\,\alpha_2) \mp \left[K^2\ell^2(1-\cos\alpha_2)^2 - 8K\tan^2\alpha_2(C_1\alpha_2^2 + 4C_2\alpha_2^2)\right]^{\frac{1}{2}}\right\} \tag{3.32}$$

Again the regions in which equilibrium is not possible are given by the inequality

$$8(C_1 + C_2)K^{-1}\ell^{-2} > (1-\cos\alpha_2)^2\alpha_2^{-2} \cot^2\alpha_2 \tag{3.33}$$

Clearly, such a forbidden region contains all sufficiently large values of α_2, in the interval $[0, \pi/2]$.

Behavior of the type illustrated in Fig. 3.8 is still possible. In fact, even more complex behavior appears to be possible for some choices of the constant $8(C_1 + C_1) K^{-1}\ell^{-2}$. It is interesting to look for the state equivalent to a multiple eigenvalue situation modelled by this 3 bar system, equating P_1 and P_2 in Eqs. 3.17 and 3.32. We have a complex non-linear system of two equations in two unknowns. A rough numerical check shows the existence of solutions for some (selected) values of the parameters. We shall not pursue detailed computations that indicate how much our system is differing from the linear model, where such prediction is fairly straight-forward.

To analyze the "multiple eigenvalue" analogue, we compare Eqs. 3.17 and 3.32, for the same value of P. We have

$$\frac{1}{2} K \cos^2\alpha_1 \{\ell (1-\cos\alpha_1) \mp [\ell^2(1-\cos^2\alpha_1)^2$$

$$- 4K^{-1}(C_1 + C_2)\alpha_1^2/\cos^2\alpha_1]^{1/2} \} = \frac{1}{4} \cot^2\alpha_2$$

$$\times \{K\ell(1-\cos\alpha_2) \mp [K^2\ell^2(1-\cos\alpha_2)^2$$

$$- 8K \tan^2\alpha_2 (C_1 + C_2)\alpha_2^2]^{1/2} \} \quad (3.34)$$

This equality is true for any values of K, ℓ, C_1, and C_2, if $\alpha_1 = \alpha_2 = 0$ and P = 0. For other values of P and for fixed values of K, ℓ, C_1, and C_2, multiple pairs $\{\alpha_1,\alpha_2\}$ exist, for which this equation is satisfied in the equilibrium (admissible) range of values of α_1 and α_2.

Our findings indicate that even large values of K, where one is liable to claim "almost rigid" bar behavior, inclusion of the compressibility effects completely changes the theoretical analysis. Without accounting for compressibility, certain phenomena that are known to occur are unexplainable within the framework of our simple mathematical model. The secret lies in the fact that the "small" terms ignored in the "usual" analysis turn out to be the leading terms (i.e. the highest powers) in the corresponding energy formulas.

An Elementary Cantilever Column Model

The "shortening" of the column of Fig. 9 is

$$\varepsilon = \ell_1(1-\cos q_1) + \ell_2 (1-\cos q_2) \quad (3.35)$$

The strain energy is

$$V = \frac{1}{2} C_1 q_1^2 + \frac{1}{2} C_2(q_2 - q_1)^2 + \frac{1}{3} C_3 (\ell_1 \sin q_1$$

$$+ \ell_2 \sin q_2)^2 \quad (3.36)$$

The total potential energy is given by

$$TPE = V - P = C_1 q_1 - C_2(q_2 - q_1) + C_3(\ell_1 \sin q_1 + \ell_2 \sin q_2)$$
$$\times \ell_1 \cos q_1 - P\,\ell_1 \sin q_1 \tag{3.37}$$

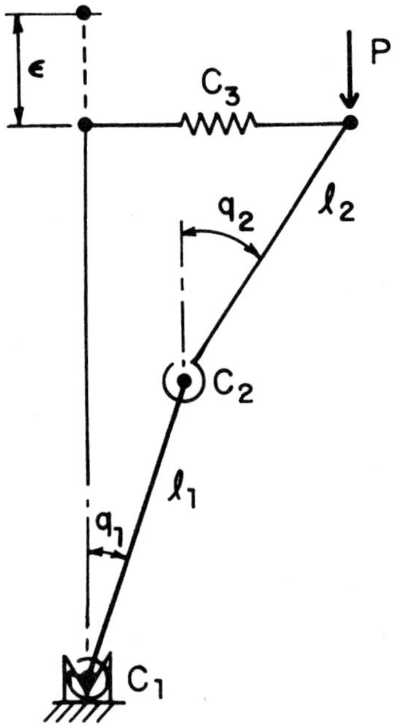

Figure 9. A Cantilever Column Model

A necessary condition for the stationary behavior of TPE is

$$\frac{\partial(TPE)}{\partial q_1} = \frac{\partial(TPE)}{\partial q_2} = 0 \tag{3.38}$$

This results in a quadratic formula for P in terms of $(C_1, C_2, C_3, \ell_1, \ell_2)$. The solutions determine stable post-buckling behavior, but the actual loss of stability can be determined by a simple linear analysis. In a straight configuration, $q_1 = 0$ $q_2 = 0$. Let q_1, q_2 be very small perturbation values from the straight configuration. The work performed by P is equal to $P\ell_1 q_1^2/2 + P\ell_2 q_2^2/2$.

The strain energy gained is

$$\frac{1}{2} C_1 q_1^2 + \frac{1}{2} C_2 q_2^2 + \frac{1}{2} C_3 (\ell_1 q_1 + \ell_2 q_2)^2 \tag{3.39}$$

where we neglect small terms of higher order.

Then, $W - V \geq 0$ if

$$P(q_1^2 \ell_1 + q_2^2 \ell_2) \geq (C_1 q_1^2 + C_2 q_2^2) + C_3(\ell_1 q_1 + \ell_2 q_2)^2$$

or

$$P \geq J = [(C_1 q_1^2 + C_2 q_2^2) + C_3(\ell_1 q_1 + \ell_2 q_2)^2] / (\ell_1 q_1^2 + \ell_2 q_2^2) \tag{3.40}$$

If $P < J$, then $q_1 = q_2 \equiv 0$ is the only stable solution. If $P > J$ then buckling occurs. $P = J$ is the critical load.

If we set $C_3 = 0$, we obtain

$$P_{crit} = (C_1 q_1^2 + C_2 q_2^2) / (\ell_1 q_1^2 + \ell_2 q_2^2) \tag{3.41}$$

In the specific case when $C = C_1 = C_2$, $P_{crit} = C/\ell$, which can be derived directly by elementary analysis.

The point made here applies fully to complex elastic structures. An elementary linear perturbation analysis suffices to predict the onset of instability, if the structure is sufficiently rigid and the unperturbed "original" stable equilibrium corresponds to the zero displacement vector. However, finding the post buckling stable positions and predicting the response to changes in design of the post buckling shape demands a complex nonlinear analysis. In particular, if trigonometric functions are approximated by finite series, then it is inadvisable to consider only the first term in such expansion for predicting the actual behavior of the structure, following the onset of instability.

Again, the effects of compressibility drastically change this analysis. Inserting telescopic springs (spring constant K) into bars of initial length ℓ_1 and ℓ_2 has such an effect. The magnitude of compression in bars of length ℓ_1 and ℓ_2 is respectively $\delta_1 = K^{-1}P/\cos q_1$

and $\delta_2 = K^{-1}P/\cos q_2$. The strain energy is

$$V = \frac{1}{2} K^{-1} P^2 \left(\frac{1}{\cos^2 q_1} + \frac{1}{\cos^2 q_2} \right) + \frac{1}{2} C_1 q_1^2$$

$$+ \frac{1}{2} C_2 q_2^2 + \frac{1}{2} C_3 (\ell_2 \sin q_2 + \ell_1 \sin q_1)^2 \tag{3.42}$$

and the work performed by the force P is

$$W = \frac{1}{2} P \{\ell_1 (1 - \cos q_1) + \ell_2 (1 - \cos q_2) + 2K^{-1}P\} \tag{3.43}$$

As before the quasi-static theory implies the equality $V = W$. We derive for P the quadratic equation

$$P^2 \left\{ \frac{1}{2} K^{-1} \left(\frac{1}{\cos^2 q_1} + \frac{1}{\cos^2 q_2} \right) - K^{-1} \right\} - \frac{1}{2} P$$

$$\times [\ell_1 (1 - \cos q_1) + \ell_2 (1 - \cos q_2)] + \frac{1}{2} (C_1 q_1^2 + C_2 q_2^2)$$

$$+ \frac{1}{2} C_3 (\ell_2 \sin q_2 + \ell_1 \sin q_1)^2 = 0 \tag{3.44}$$

This we solve for P and again find the regions in which stability is impossible, defining the so-called snap-through regions. Clearly, stability is impossible (in the region $0 \leq q_1 \leq \pi/2$, $0 \leq q_2 \leq \pi/2$) if either one of the angular displacements q_1 or q_2 is large. Again, detailed analysis confirms patterns of behavior that have no counterpart in the linear theory (if one assumes q_1 and q_2 are "small") or even in the elastic theory if one assumes (as Euler did) that the length of each bar does not change.

All problems discussed above are accentuated if one considers the optimal design of a structure, the optimality criterion being the force causing onset of instability for a given (constant) weight, or the weight of the structure for a given critical load. For examples of such problems see [4], [5], [6], [7], [13], and [14].

Examples of simple finite element structures that simulate roughly the behavior of a buckled column, such as those illustrated in Figs. 2, 3, 4, 5, and 6 reveal some essential difficulties that must be faced if a realistic model of column buckling is to be found. In what follows, we extend our analysis to a slightly modified model, which is in some way close to the Euler model. We continue to utilize the quasi-static approach.

4. FORMULATION OF A NONLINEAR BUCKLING PROBLEM FOR A SLENDER COLUMN

In what follows we assume Hooke's law for the material, but some nonlinear geometric terms will be retained. We shall also retain the terms representing purely compressive effects, aside from the bending effect of the load applied. The process by which the column assumes its buckled shape can be analyzed using either the quasi-static or the dynamic analysis. If one considers only stable equilibrium positions, kinetic energy effects can be ignored. They cannot be ignored if one wishes to understand the actual mechanism of buckling behavior. In this section, we proceed by the so-called quasi-static procedure, ignoring the kinetic energy terms. Shear effects are still ignored here. We predict that inclusion of shear will have similar effect on our analysis as the inclusion of direct compression.

Let us consider the virtual work performed by the load P in deflecting the column.

The process of buckling the column is visualized in three steps, including compressibility effects, as shown in Fig. 10. As the force P is increased to the critical value, the column is compressed to length $\bar{\ell}$ but retains the shape $y = y(x) = 0$. At the value $P = P_{crit}$, the shape changes to a shape $y = y(x; P)$. Two separate effects take place. The length of column is decreased by the amount δ_G given by

$$\delta_G = \int_0^{\bar{\ell}} \left(\frac{\theta^2}{2!} - \frac{\theta^4}{4!} + \dots \right) ds , \tag{4.1}$$

where θ denotes the angle between the axis of the column and the original axis of the undeflected column and $\bar{\ell}$ denotes the compressed length of the column, prior to the change of shape. The second effect is an increase in the length of the column, because the force P does not act along the column's axis. The amount of axial "lengthing" is given by

$$\delta_c = \int_0^{\bar{\ell}} P \frac{(\cos^2\theta - \cos\theta)}{EA(s)} ds \tag{4.2}$$

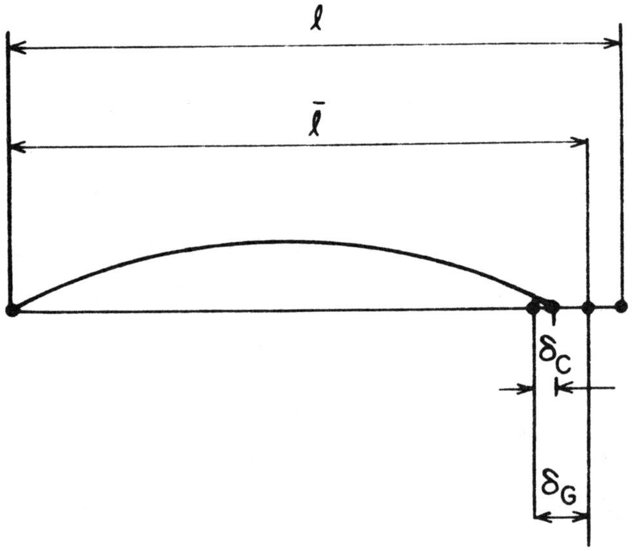

Figure 10. Deflection of A Continuous Column.

The total deflection is given by

$$\delta = \delta_G + \delta_c = \int_0^{\bar{\ell}} (\theta^2/2! - \theta^4/4! + \ldots)\, ds$$

$$- \int_0^{\bar{\ell}} \frac{P}{EA(s)} (\theta^2/2 + \frac{7}{24}\theta^4 + \ldots)\, ds \tag{4.3}$$

The second term in this expression is usually ignored, except for very slender beams, where $\int_0^{\bar{\ell}} P/EA(s)\, ds$ is comparable in magnitude to the term δ_G. This situation does occur in certain design optimization problems.

The strain energy of the deflected column is given by

$$SE = \frac{P^2}{2}\int_0^{\bar{\ell}} \frac{\cos^2\theta(s)}{2EA(s)}\,ds + \frac{1}{2}\int_0^{\bar{\ell}} EI(s)(\theta')^2 ds \tag{4.4}$$

The increase in strain energy is given by

$$\Delta SE = \frac{1}{2}\int_0^{\bar{\ell}} EI(s)(\theta')^2 ds - \frac{P^2}{2}\int_0^{\bar{\ell}} [(\theta^2 - \theta^4/3 + \ldots)/EA(s)]ds \tag{4.5}$$

We comment that while fourth powers of $\theta(s)$ occur in the integrals, it suffices to assume that $\theta \in H_0^1(\bar{\ell})$. The existence of the integral of $(\theta')^2$ implies the existence of the integral of θ^4, and the problems of column stability can be considered within the framework of the Sobolev space $H_0^1(\bar{\ell})$.

We assume throughout that θ is not "large", in the sense that the expansion given in Eqs. 4.1 to 4.5 can be truncated after terms of fourth order; i.e., terms of order six and higher can be neglected. This implies that the elastica theory need not be invoked. Roughly speaking, we do not assume that deflections are "small", but we still shy away from very large deflections. To formulate stability theory we compute the change in total potential energy between the original (unbuckled) state $\theta(s) \equiv 0$ and the state given by $\tilde{\theta}(s)$, where $\tilde{\theta}(s) \not\equiv 0$,

$$\begin{aligned}\Delta V &= \Delta SE - P\delta \\ &= \frac{1}{2}\int_0^{\bar{\ell}} EI(s)(\theta')^2 ds - \int_0^{\bar{\ell}} P(\theta^2/2 - \theta^4/24 + \ldots)ds \\ &\quad - \frac{1}{2}\int_0^{\bar{\ell}} P^2\,(11/12\theta^4 - \ldots)\,ds\end{aligned} \tag{4.6}$$

The minimum potential energy principle can be stated as the following lemma, which is valid for the quasi-static and for the dynamic theory.

LEMMA 1. The trivial solution $\theta(s) \equiv 0$ is unstable if and only if there exists a function $\tilde{\theta}(s) \not\equiv 0$ such that $\Delta V(\tilde{\theta}(s)) < 0$, where $\Delta V(\tilde{\theta}(s))$ is given by Eq. 4.6.

The inequality $\Delta V(\tilde{\theta}) < 0$ written out in full is a quadratic inequality in P ; i.e.,

$$P^2 \int_0^{\bar{\ell}} \left[(EA(s))^{-1}\ 11/24\ \tilde{\theta}^4(s)\right] ds$$

$$+ \quad P\int_0^{\bar{\ell}} (\tilde{\theta}^2/2 - \tilde{\theta}^4/24)\, ds - 1/2\int_0^{\bar{\ell}} EI(s)(\tilde{\theta}')^2 ds \geq 0 \tag{4.7}$$

As before, powers of order six and higher have been neglected. Since only positive values of P are considered in buckling theory, the inequality of Eq. 4.7 can be written as

$$P \geq \left[11/12 \int_0^{\bar{\ell}} \frac{\tilde{\theta}^4}{EA(s)}\, ds \right]^{-1} \Bigg\{ -\int_0^{\bar{\ell}} (\tilde{\theta}^2/2 - \tilde{\theta}^4/24)\, ds$$

$$+ \left[\left(\int_0^{\bar{\ell}} (\tilde{\theta}^2/2 - \tilde{\theta}^4/24)\, ds \right)^2 + \left(\int_0^{\bar{\ell}} \frac{11/12\tilde{\theta}^4}{EA(s)}\, ds \right)\right.$$

$$\left. \times \left(\int_0^{\bar{\ell}} EI(s)(\tilde{\theta}')^2\, ds \right) \right]^{1/2} \Bigg\} = F(\tilde{\theta}) \tag{4.8}$$

We shall refer to the problem posed in equations and inequalities of Eqs. 4.3 to 4.8 as the nonlinear column problem. We refer to a pair $\{\tilde{\theta}(s), \tilde{P}\}$ as a critical pair if $\tilde{\theta}(s) \not\equiv 0$, the function $\tilde{\theta}(s)$ is a continuous function, $\tilde{\theta}(s) \in H_0^1(\bar{\ell})$, and $F(\tilde{\theta})$ attains an extremal value, while $\tilde{P}$ is equal to $F(\tilde{\theta})$.

<u>THEOREM 1</u>. Suppose that $F(\tilde{\theta})$ is Fréchet differentiable. Then for small values of θ and very small values of $[EA(s)]^{-1}$, as compared to EI(s), the critical value of $\tilde{P}$ given by Eq. 4.8 is close the the value of the Rayleigh quotient. We mean by this statement that

$$\lim_{A^{-1}\to 0} F(\theta) = \phi(\bar{\ell}, \theta', \theta^2, \theta^4) \text{ and } \lim_{\theta^4\to 0} \phi(\bar{\ell}, \theta', \theta^2, \theta^4) = \mathcal{R}$$

where $\mathcal{R}$ denotes the Rayleigh quotient associated with the linear problem of Eq. 1.1.

Proof: We denote by

$$\left.\begin{aligned} \mathcal{A} &= 11/12 \int_0^{\bar{\ell}} \theta^4/EA(s)\, ds \\ \mathcal{B} &= \int_0^{\bar{\ell}} (\theta^2/2! - \theta^4/4!)\, ds \\ \mathcal{C} &= \int_0^{\bar{\ell}} EI(s)(\theta')^2\, ds \end{aligned}\right\} \tag{4.9}$$

If $F(\theta)$ is Fréchet differentiable, the derivative $F'(\theta)$ is given by

$$F'(\theta) = \mathcal{A}^{-1} \Big\{\Big[- \partial\mathcal{B}/\partial\theta + 1/2\, (\mathcal{B}^2 + \mathcal{AC})^{-1/2}$$

$$\times (2\mathcal{B}\, \partial\mathcal{B}/\partial\theta) + \partial\mathcal{A}/\partial\theta\, \mathcal{C} + \mathcal{A}\, \partial\mathcal{C}/\partial\theta\Big] - F(\theta)\, \partial\mathcal{A}/\partial\theta\Big\}$$

$F'(\tilde{\theta}) = 0$ is a necessary condition for the pair $\{\tilde{\theta},\tilde{P}\}$ to be the critical pair, provided $F'(\tilde{\theta})$ exists.

$$\frac{\partial\mathcal{A}}{\partial\theta} = \frac{\pi}{3}\, \frac{\theta^3}{EA(s)}$$

$$\frac{\partial\mathcal{B}}{\partial\theta} = \theta - \frac{1}{6}\, \theta^3$$

$$\frac{\partial\mathcal{C}}{\partial\theta} = 2(EI(s)\theta')'$$

The relation $F'(\theta) = 0$ becomes

$$F'(\theta) = \left(- \hat{\theta} + \frac{1}{6}\, \hat{\theta}^3\right) + \frac{1}{2}\left[\left(\int_0^{\bar{\ell}} \left(\frac{\hat{\theta}^2}{2} - \frac{\hat{\theta}^4}{24}\right) ds\right)^2\right.$$

$$\left. + \left(\frac{11}{12} \int_0^{\bar{\ell}} \frac{\hat{\theta}^4}{EA(s)}\, ds\right)\left(\int_0^{\bar{\ell}} EI(s)(\hat{\theta}')^2\, ds\right)\right]^{-1/2}$$

$$\times \left[2\int_0^{\bar{\ell}} \left(\frac{\hat{\theta}^2}{2} - \frac{\hat{\theta}^4}{24}\right) ds \left(\hat{\theta} - \frac{1}{6}\, \hat{\theta}^3\right) + \frac{11}{3}\, \frac{\hat{\theta}^3}{EA(s)}\left(\int_0^{\bar{\ell}} EI(s)(\hat{\theta}')^2 ds\right)\right.$$

$$\left. - \frac{11}{6}\, (EI(s)\hat{\theta}')'\left(\int_0^{\bar{\ell}} \frac{\hat{\theta}^4}{EA(s)}\, ds\right)\right] - \frac{11}{3}\, \hat{\theta}^3\, (EA)^{-1}\, F(\hat{\theta}) = 0 \;. \tag{4.10}$$

For small values of $\hat{\theta}^4(EA(s))^{-1}$, after use of L'Hospital's rule, we obtain

$$P_{crit} = \frac{\int_0^{\bar{\ell}} EI(\hat{\theta}')^2\, ds}{\int_0^{\bar{\ell}} \left(\hat{\theta}^2 - \frac{\hat{\theta}^4}{12}\right) ds} \approx \frac{\int_0^{\bar{\ell}} EI(s)(\hat{\theta}'(s))^2\, ds}{\int_0^{\bar{\ell}} \hat{\theta}^2\, ds}$$

which is exactly the Rayleigh quotient, with the extremal function describing the deflection of the column. This completes the proof.

We have omitted the details of some rather cumbersome computations. The basic formula

$$\lim_{a\to 0} \left(\frac{-b+\sqrt{b^2 - 4ac}}{2a}\right) = \frac{c}{b}$$

was used in applying L'Hospital's rule and the relationship

$$0 = \int_0^{\bar{\ell}} (\theta - \theta^3/3!)\, ds \approx \int_0^{\bar{\ell}} \sin\theta\, (s)\, ds \tag{4.12}$$

was used in simplifying some relations. We comment that the usually accepted form of Rayleigh's quotient is derived if the terms in θ of fourth and higher powers are ignored and θ is approximated by $y'(x)$.

<u>THEOREM 2</u>. Within approximation of the order of $\int_0^{\bar{\ell}} \theta^6(s)ds$, the value of P_{crit} given by Eq. 4.10 is always lower than the one predicted by the Rayleigh quotient $\mathcal{R}$.

<u>Proof</u>: We solve Eq. 4.10 for $F(\theta)$ and use the simple support condition $\hat{\theta}'(0) = \hat{\theta}'(\ell)$ in integrating by parts to derive

$$F(\theta') = \left[\left(\int_0^{\bar{\ell}} \left(\frac{\hat{\theta}^2}{2!} - \frac{\hat{\theta}^4}{4!}\right) ds\right) + \frac{11}{12}\left(\int_0^{\bar{\ell}} EI(\hat{\theta}')^2 ds\right) \times \left(\int_0^{\bar{\ell}} \frac{\hat{\theta}^4}{EA(s)}\, ds\right)\right]^{-\frac{1}{2}} \frac{1}{2}\int_0^{\bar{\ell}} EI(\hat{\theta}')^2 ds \tag{4.13}$$

This value is clearly smaller than the one given by Eq. 4.11, since the numerators of both expressions are identical, but Eq. 4.13 contains an additional positive term in the denominator. This completes the proof.

This theorem has some important physical consequences. In particular if one loads and unloads a column past the critical load, buckling will occur at a higher load than the recovery to the originally straight shape. In the onset of buckling, only the quadratic terms need to be considered, because $\theta(s)$ is extremely small everywhere and higher order terms do not contribute significantly to the onset of instability. The higher order terms may contribute noticeably to the recovery, when the load is slowly reduced to a value smaller than the Euler critical value.

The inclusion of the so called "small" compressive term and the retention of terms of fourth order have produced unexpected theoretical results. These results better reflect the actual behavior of elastic columns. The equations modelling the buckled shape of a column can be solved for any value of the load P (within reason that is, since for very large values of P the elastic behavior, which is tacitly assumed here is no longer physically acceptable). This theory also predicts the experimentally observed phenomenon that the buckling load and the load at which the column "snaps back" to a straight shape differ slightly, with the original buckling load being slightly higher. Moreover, this theory predicts irreversibility and the necessity of considering dissipative terms in the mathematical model.

Differentiability of the critical load

If $\hat{\theta}(s)$ is unique, then a unique derivative exists, given by Eqs. 4.9 and 4.10. Let us suppose that shapes $\hat{\theta}_1$ and $\hat{\theta}_2$ exist such that

$$F(\hat{\theta}_1) = F(\hat{\theta}_2) \leq F(\theta) \qquad (4.14)$$

for any $\theta(s) \neq \theta_1(s)$, $\theta(s) \neq \theta_2(s)$. Let us further suppose that the Fréchet derivative of the functional F exists in some neighborhood of $\hat{\theta}_1$, and of $\hat{\theta}_2$. Computing these derivatives and equating them to zero we have the following set of equalities.

$$\left.\frac{\partial P_{crit}}{\partial \theta}\right|_{\hat\theta=\hat\theta_1} = 2 \left\{\left[(EI(s)\hat\theta')' - P_{crit}(\hat\theta - \hat\theta^3/3!)\right]\right.$$

$$\times \left.\int_0^{\bar\ell} (\hat\theta^2 - \hat\theta^4/12)ds\right|_{\hat\theta=\hat\theta_1}$$

$$= 2 \left.\left\{(EI(s)\hat\theta')' - P_{crit}(\hat\theta - \hat\theta^3/3!)\right\} / \mathcal{B}(\hat\theta)\right|_{\hat\theta=\hat\theta_1} \qquad (4.15)$$

That is, $\hat\theta_1$ satisfies the nonlinear eigenvalue equation of Eq. 4.15. Clearly, $\hat\theta_2$ satisfies the same equation and both $\hat\theta_1$ and $\hat\theta_2$ satisfy boundary conditions $\theta_1'(0) = \theta_1'(\ell) = 0 = \theta_2'(0) = \theta_2'(\ell)$ and $\int_0^\ell \hat\theta(s)ds = 0$.

Since $\hat\theta_1$ and $\hat\theta_2$ provide a local minima for P_{crit}, we can equate the right hand side of Eq. 4.15 to zero and similarly treat the identical equation for $\hat\theta_2$. Hence, we have

$$(EI(s)\hat\theta_i')' - P_{crit}(\hat\theta_1 - \hat\theta_1^3/3!) = 0 \ , \ i = 1,2 \qquad (4.16)$$

That is, the nonlinear eigenvalue problem of Eq. 4.16 must have a multiple eigenvalue at $P = P_{crit}$. If we identify P_{crit} with P_{Euler}, we obtain a contradiction. Computing the derivative of the lowest eigenvalue of the linear problem, that is the derivative of the Euler buckling load, and legitimately ignoring terms of order higher than two, we conclude that

$$P_{Euler}\hat\theta_1 = (-EI(s)\hat\theta_1')',$$

$$P_{Euler}\hat\theta_2 = (-EI(s)\hat\theta_2')', \text{ and}$$

$$\frac{\int_0^{\bar\ell} EI(s)(\hat\theta_1)^2 ds}{\int_0^{\bar\ell} \hat\theta_1^2 ds} = \frac{\int_0^{\bar\ell} EI(s)(\hat\theta_2')^2 ds}{\int_0^{\bar\ell} \hat\theta_2^2 ds} .$$

In effect, we have too many relations if we also invoke the equality of finite potential energy; i.e., Eq. 4.15. Since $\hat\theta_1 = -\frac{1}{P_{Euler}}(EI\hat\theta_1')'$, with the boundary conditions and $\hat{\theta}_2 = -\frac{1}{P_{Euler}} (EI\,\hat{\theta}_2')'$ with the same conditions, it follows that $\hat{\theta}_1$ and $\hat{\theta}_2$ are equal. This result can be summarized in the following manner.

THEOREM 3. P_{Euler} is a differentiable function of $\hat{\theta}$ (in the Fréchet sense) if and only if it is a simple eigenvalue of the corresponding linear problem of Eq. 1.1.

THEOREM 4. The functional P_{Euler} is a differentiable function of the design parameter I(s) if and only if P_{Euler} is a simple eigenvalue of the linear problem of Eq. 1.1.

Proof. The technique of proof is identical with that of the preceeding theorem (also see the paper of E. J. Haug and B. Rousselet [15]).

THEOREM 5. Let the loads and the deflections be bounded. There exists a continuous range of values $P = P_{crit}$, within such a bound that the nonlinear problem of column deflection has a solution.

Proof: We only need to show that, given any $P \geq P_{crit}$, there exists a $\hat{\theta}(s) \epsilon H_0^1[0_1\bar{\ell}]$ such that $V(\hat{\theta}(s))$ attains a minimum. Since for any admissible displacements $\theta(s)$, $0 \leq s \leq \bar{\ell}$, $V(\hat{\theta}(s))$ has a lower bound, it is not hard to show that for the given $P \geq P_{crit}$ it has the greatest lower bound $\tilde{V}$. Let us select a sequence of $H_0^2(\bar{\ell})$ admissible functions $\theta_i(s)$ such that $\lim_{i\to\infty} V(\theta_i) = \tilde{V}$, which exists by a physical consideration. Since the map $H_0^2(\bar{\ell}) \to H_0^1(\bar{\ell})$ is compact and since the moments and shears are uniformly bounded, there exists $\hat{\theta} \epsilon H_0^1[0,\bar{\ell}]$ such that $V(\hat{\theta}(s)) = \tilde{V}$. This argument does not imply anything about the uniqueness of $\hat{\theta}(s)$. What it does imply is physically sensible, but it directly contradicts the linear eigenvalue theory based on "the neglect of small terms" argument.

5. IRREVERSIBILITY OF THE NONLINEAR COLUMN BUCKLING PROCESSES

In the classical theory of beams and columns, elasticity implies reversibility. The technical term used for such processes, if the kinetic terms are neglected, is quasi-static (reversible) elastic processes. Even if the kinetic energy is considered in the dynamic theory of column buckling, it is assumed that the kinetic energy, which is gained during the buckling deformation, is fully converted into the potential energy of the deformed column.

Let us consider the load that coincides with the first eigenvalue of the linear problem of Eq. 1.1. We denote its value by P_{Euler}. The work performed by the load, as the column deflects from $\delta = 0$ to $\delta = \bar{\delta}$, is given by

$$\int_0^{\delta = \delta} P(\delta)d\delta \tag{5.1}$$

with

$$P(0) = P_{Euler}$$

$$P(\bar{\delta}) = P_{crit} < P_{Euler} ,$$

where P_{crit} is the (nonlinear) critical load obtained by asserting strict equality in Eq. 4.8.

<u>LEMMA 3</u>. The buckling process is fully reversible if and only if the following equality holds:

$$\int_0^{\delta = \bar{\delta}} P(\delta)d\,\delta = 1/2 \int_0^{\bar{\ell}} EI(s)(\theta ')^2 ds$$

$$- P_{crit} \int_0^{\bar{\ell}} \left[\left(\hat{\theta}^2 - \hat{\theta}^4/3 + \ldots\right) / \left(EA(s)\right)\right] ds \tag{5.2}$$

<u>Proof</u>: We simply confirm the statement that the work performed is converted into potential energy. That is, whatever kinetic energy is developed is converted into a finite value of the potential energy. This corresponds exactly to the Eq. 5.2.

<u>REMARK</u>: The buckling force cannot remain constant during the buckling process.

<u>Proof</u>: Suppose P = constant, then $P = P_{crit} = P_{Euler}$ = constant. This contradicts our previously proved inequality $P_{crit} < P_{Euler}$.

THEOREM 6. During the reversible buckling process, $P(\delta)$ is a monotone function of δ.

Proof: Let us suppose that on some small interval (α,β), P has an almost stationary behavior, $P \approx \bar{P}$. This can be restated with sufficient number of ε-s and δ-s thrown in, but this statement should be clear. Then,

$$P^2 \int_0^{\bar{\ell}} [11/12(\theta_\alpha^4(s) - \theta_\beta^4(s)) / (EA(s))]ds$$

$$+ \bar{P} \int_0^{\bar{\ell}} [(\theta_\alpha^2(s) - \theta_\beta^2(s)) - (\theta_\alpha^4(s) - \theta_\beta^4(s))]\, ds$$

$$- \int_0^{\bar{\ell}} EI(s)\left[\left(\frac{d\theta_\alpha}{ds}\right)^2 - \left(\frac{d\theta_\beta}{ds}\right)^2\right] ds = 0$$

This can be reduced to the form

$$A^{-1}(\theta_\alpha - \theta_\beta)\{B(\theta_\alpha - \theta_\beta) + (B(\theta_\alpha - \theta_\beta)$$

$$+ AC(\theta_\alpha - \theta_\beta))^{1/2}\} = 1/2 \quad (B^2(\theta_\alpha - \theta_\beta)$$

$$+ AC(\theta_\alpha - \theta_\beta))^{-1/2}\, C\,(\theta_\alpha - \theta_\beta) \qquad (5.3)$$

Here the bracketed quantity $(\theta_\alpha - \theta_\beta)$ indicates the difference in values when the function is computed first at the value corresponding to the shape $\theta_\alpha = \theta_1(s)$, at the deflection point $\delta = \alpha$, and then at the shape $\theta_\beta(s) = \theta_2(s)$, at the deflection point $\delta = \beta$. Equation 5.3 indicates that one can solve this relation for $B(\theta_\alpha - \theta_\beta)$ in terms of $A(\theta_\alpha - \theta_\beta)$ and $C(\theta_\alpha - \theta_\beta)$. But this is, in general, impossible, since B contains no design parameters, while A and C depend on $A(s)$ and $I(s)$.

Our arguments indicate that it is difficult to produce a reversible buckling process. The difficulty lies in converting kinetic energy into potential energy. Our present model does not provide a mechanism for such conversion. Since dissipative terms were not included either in our scheme, our model does not provide any mechanism for either generating or disposing of kinetic energy terms. Specifically, we have pursued the so-called quasi-static approach. One has to witness only a single buckling experiment to convince himself that the process is certainly not quasi-static. The buckling of a column occurs with an audible "snap" and the shape changes with a visible high speed and subsequent slight tremor accompanying the postbuckled shape. Fortunately, in most designs, the postbuckled shape represents a stable equilibrium state and the "tremor" represents a dissipation of the excess kinetic energy, before settling to the stable postbuckled shape.

6. STABILITY OF THE POSTBUCKLED SHAPE

The postbuckled shape is determined by seeking the minimum of the potential energy for a given load $P \geq P_{Euler}$. The shape of the postbuckled column must satisfy the condition of stationary behavior of the potential energy. We consider the first variation of the potential energy, which is given by

$$\begin{aligned} -\delta V = 0 = \{ & P^2 (EA(s))^{-1}\, 11/6\, \hat{\theta}^3 \\ & + P(\hat{\theta} - \hat{\theta}^3/3!) + (EI(s)\theta')'\}\, \delta\theta \\ & + \{2P \int_0^{\bar{\ell}} \left((EA(s))^{-1} \cdot 11/24\, \theta^4(s)\right) ds \\ & + \int_0^{\bar{\ell}} (\theta^2/2 - \theta^4/24) ds\}\, \delta P \end{aligned} \tag{6.1}$$

The condition of stable equilibrium is provided by the positive definite property of the Hessian matrix

$$\begin{pmatrix} \dfrac{\partial^2 V}{\partial \theta^2} & \dfrac{\partial^2 V}{\partial \theta \partial P} \\ \dfrac{\partial^2 V}{\partial P \partial \theta} & \dfrac{\partial^2 V}{\partial P^2} \end{pmatrix} \tag{6.2}$$

Formal computation gives the matrix of operators $H_0^1 \to H_0^1$

$$\begin{bmatrix} \{11/2\ P^2 (EA)^{-1} \theta^2 + P(1-\theta^2/2) + d/ds\ (EI\ d/ds)\} , & V_{\theta P} \\ V_{P\theta} & , 11/24 \int ((EA)^{-1\ 4}/4!)\, ds \end{bmatrix}$$

where $V_{\theta P} = 2P\,((EA)^{-1}\ 11/6\ \theta^3) + (\theta - \theta^3/3!)$.

This turns out to be positive definite if terms including $(EA)^{-1}$ are small; i.e., if the compressive effects are small compared to the bending energy terms. However, this is not necessarily the case in optimum designs, which may include very slender sections, such as discussed in [8], [9], or [11]. In such cases, buckling may truly correspond to a catastrophic collapse of a structure. This aspect of collapse of a structure, and other possibilities of catastrophic collapse have only been touched upon in the existing literature, with perhaps the best known example of a complex structure prone to a catastrophic collapse being given by the Mitchell truss.

7. CONCLUDING REMARKS

The analysis was carried out in this paper specifically for the buckling of an elastic column. However, the inclusion of compressive effects and retention of some higher order terms may be applied to other problems, such as the stability in bending of thin plates that are subjected to edge loads. The operator Ly=(EIy")" has to be replaced by the corresponding operator.

$$Lw = \left[h^3 w_{xx} + \nu w_{yy}\right]_{xx} + \left[h^3 w_{yy} + \nu w_{xx}\right]_{yy}$$
$$+ 2(1-\nu)(h^3 w_{xy})_{xy}$$

but the basic problems and stability argument remain unchanged.

It would be of interest to investigate how a more general approach would affect the optimal design results of [13] or [14]. Further extensions to optimization of shell design and to complex structural design can be visualized along the same lines.

The inclusion of compression strain energy terms permitted us to carry out a detailed quasi-static analysis of a buckled column and of very simple finite element structures consisting of only two or three elements. Our analysis indicated effects that cannot be predicted by Euler's (linear) theory. We still ignored the effects of shear, which can be taken care of by adding the term

$$1/2 \int_0^{\bar{\ell}} \left[G\alpha A(x)(W_x)^2\right]dx$$

to the strain energy, where $W(x)$ denotes the transverse displacement and G is a constant depending on E and ν. The kinetic energy terms including either

$$T = 1/2 \int_0^{\bar{\ell}} \left[\hat{\rho} A(x)(W_t)^2\right]dx$$

or a more exact term

$$1/2 \int_0^{\bar{\ell}} \left[\hat{\rho} A(x) \left((W_t)^2 + r_g (W_{xt})^2\right)\right] dx$$

which should be included in a more exact analysis of the buckling behavior. Here, r_g is the radius of gyration of the cross-sectional area about the neutral axis and $\hat{\rho}$ is the mass density of the material. While no experimental confirmation is needed for the three and two bar models (except to confirm that no serious error was made in the computation), experiments on very slender shapes will be of help to indicate the importance of compression, shear, and of the dynamic effects and to provide the missing link between our nonlinear theories and Euler's "elastica" theory.

REFERENCES

[1] V. Komkov, and E.J. Haug, "The shape of an optimal column," Proceedings of Conference on Optimization of Distributed Parameter Structures, Iowa City, Iowa, 1980, E.J. Haug and J. Cea, editors, Sijthhoff and Noordhoff Publishers, Alphen ann den Rijn, Holland, to appear in 1981.

[2] C. Truesdell, The rational mechanics of flexible or elastic bodies 1638-1788. Introduction to Leonhardi Euleri Opera Omnia, vol. X, sieriei secundae. Society for the Natural Sciences of Switzerland, 1960, p. 343-370.

[3] J. Den Hartog, Advanced Strength of Materials, McGraw-Hill, New York, 1957.

[4] N. Olhoff and S.H. Rasmussen, "On single and bimodal buckling loads of clamped columns," Int. J. Solids Struct. 13, 1977, p. 605-614.

[5] N. Olhoff and F.I. Niordson, "Some problems concerning singularities of optimal beams and columns," ZAMM, 59, 1979.

[6] N. Olhoff, "Optimal design against structural vibration and instability," Ph.D. thesis, Technical University of Denmark, Lyngby, Denmark, 1978.

[7] N. Olhoff and J.E. Taylor, "On optimal structural remodelling," J. Optimization Theory and Appl. vol 27, 1979, p. 571-582.

[8] J.B. Keller, "The shape of the strongest column," Arch. Rat. Mech. Anal. 5, 1960, p, 275-285.

[9] I. Tadjbaksh and J.B. Keller, "Strongest columns and isoperimetric inequalities for eigenvalues," J. Appl. Mech., 9, 1962, p. 159-164.

[10] Souhail El-Asfouri, O. Johnson and B. McInnis, "On the problem of minimizing the volume of a vibrating cantilever," Quat. Appl. Math., July, 1976, p. 203-206.

[11] F.I. Niordson, "The optimal design of a vibrating beam," Quat. Appl. Math., 23. 1965m p. 47-53.

[12] K.K. Choi and E.J. Haug, "Optimization of Structures with Repeated Eigenvalues", Optimization of Distributed Parameter Strutures, E.J. Haug and J. Cea, Editors, Sijthoff and Noordhoff Publishers, Alphen ann den Rijn, Holland, 1981.

[13] N.V. Banichuk, "Design of a plate for minimum deflection and stress," Proceedings of the Conference on Optimization of Distributed Parameter Structures, Iowa City, Iowa, May 1980. E.J. Haug and J. Cea, Editors, Sijthoff and Noordhoff Publishers, Alphen ann den Rijn, Holland, to appear in 1981.

[14] Ken Tung Cheng and N. Olhoff, "An investigation concerning the optimal design of solid elastic plates," Report 174, March 1980, The Danish Center for Applied Mathematics and Mechanics, The Technical University of Denmark, 2800, Lyngby, Denmark.

[15] E.J. Haug, and B. Rousselet, "Design Sensitivity Analysis In Structural Mechanics II, Eigenvalue Variations," Journal of Structural Mechanics, Vol 8, No. 2, 1980, p. 161-186.

VADIM KOMKOV
DEPARTMENT OF MATHEMATICS
COLLEGE OF ARTS AND SCIENCES
WEST VIRGINIA UNIVERSITY
MORGANTOWN, WV 26506

E. J. HAUG
DIVISION OF MATERIALS ENGINEERING
COLLEGE OF ENGINEERING
THE UNIVERSITY OF IOWA
IOWA CITY, IOWA 52242

Contemporary Mathematics
Volume 4
1981

VIBRATIONS AND STABILITY OF SHALLOW ELASTIC ARCHES

Raymond H. Plaut[1]

ABSTRACT. Shallow elastic arches with pinned or clamped ends are considered. Some results dealing with the vibrations and stability of these arches are reviewed. For static loading, Liapunov's direct method is applied to determine conditions for stability and instability of equilibrium states. Linear vibrations about an equilibrium configuration are examined next, followed by a discussion of the response to dynamic loads. Finally, the influence of load position on the critical load and the interaction effects of multiple, independent loads are investigated.

1. INTRODUCTION. Shallow elastic arches are typical of a class of structures which exhibit prebuckling deformations and snap-through instability. They have been studied extensively, in part due to the simple form of the governing equations. This paper describes a few of these studies and presents some illustrative examples. An earlier, similar review was presented in Ref. [1].

The equation of motion is listed in Section 2. For a pinned, sinusoidal arch under sinusoidal static loading, the equilibrium states are given in Section 3. In Section 4, Liapunov's direct method is used to investigate the stability of these states. Instability occurs at either a limit point or a bifurcation point. Vibration modes and frequencies are determined in Section 5 for small motions about an equilibrium state.

In Section 6, dynamic loads are considered. A definition of critical load is given, and a sufficient stability condition is presented. Also, an upper bound on the response is obtained, and the application of Galerkin's method to compute numerical solutions is discussed.

A concentrated load located at an arbitrary position along the arch is considered in Section 7. Some plots of critical load versus load position are presented, with the load applied either quasi-statically, stepwise (in time), or impulsively. In the static case, the equilibrium configuration

---

1980 Mathematics Subject Classification, 73H05.

[1] Supported by the National Science Foundation under Grant No. CME-7920781.

is determined, and the imperfection-sensitivity at bifurcation points is examined by means of asymptotic expansions.

Finally, in Section 8, independent loads are placed at the third-points of the arch, and interaction curves of critical load combinations are plotted in the loading plane. The loads are applied quasi-statically or stepwise. If the arch rise is sufficiently large, so that bifurcation instability occurs when the static loads are equal in magnitude, then these stability boundaries are quite different in form compared to those for shallower arches which can only permit limit point instability.

2. FORMULATION. Consider the shallow elastic arch shown in Fig. 1. It is homogeneous and has uniform cross-section. The ends are pinned at a distance L apart. Initially, the coordinate of the centerline of the arch (dash-dot line) is $Y_o(X)$. A distributed load Q(X) is applied quasi-statically to the arch, and the resulting equilibrium configuration is $Y_s(X)$. (Concentrated loads can be included with the use of Dirac delta functions.) A dynamic load P(X,T) may then be applied, beginning at time T = 0.

If Y(X,T) denotes the configuration of the arch at time T, the equation of motion is assumed to be [2,3]

$$\mu \frac{\partial^2 Y}{\partial T^2} + EI \frac{\partial^4 (Y-Y_o)}{\partial X^4} + M \frac{\partial^2 Y}{\partial X^2} = - Q - P \tag{1}$$

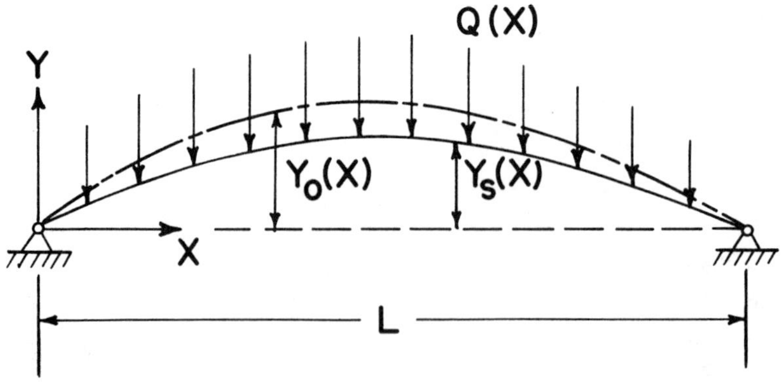

Fig. 1. Geometry of Arch

where

$$M = \frac{EA}{2L} \int_0^L \left[\left(\frac{\partial Y_o}{\partial X} \right)^2 - \left(\frac{\partial Y}{\partial X} \right)^2 \right] dX, \tag{2}$$

μ is the mass per unit length, E is Young's modulus, I is the moment of inertia of the cross-section, and A is the cross-sectional area. The boundary conditions are given by

$$Y = 0, \; EI \frac{\partial^2 (Y - Y_o)}{\partial X^2} = 0 \text{ at } X = 0, L. \tag{3}$$

Consider the dimensionless quantities

$$x = \frac{X}{L}, \; y = \frac{Y}{2} \sqrt{\frac{A}{I}}, \quad y_o = \frac{Y_o}{2} \sqrt{\frac{A}{I}}, \; y_s = \frac{Y_s}{2} \sqrt{\frac{A}{I}},$$

$$t = T \sqrt{\frac{EI}{\mu L^4}}, \quad q = \frac{QL^4}{2\pi^4 EI} \sqrt{\frac{A}{I}}, \; p = \frac{PL^4}{2\pi^4 EI} \sqrt{\frac{A}{I}}. \tag{4}$$

Let dots and primes denote partial differentiation with respect to t and x, respectively. Then (1) becomes

$$\ddot{y} + y'''' - y_o'''' + \eta\, y'' = - \pi^4 q - \pi^4 p \tag{5}$$

where

$$\eta = 2 \int_0^1 [(y_o')^2 - (y')^2] dx, \tag{6}$$

and the boundary conditions take the form

$$y = 0, \; y'' - y_o'' = 0 \quad \text{at } x = 0, 1. \tag{7}$$

From (5), note that the dimensionless equilibrium configuration $y_s(x)$ satisfies the equation

$$y_s'''' - y_o'''' + \gamma^2 y_s'' = - \pi^4 q \tag{8}$$

where

$$\gamma^2 = 2 \int_0^1 [(y_o')^2 - (y_s')^2] dx. \tag{9}$$

The quantity γ^2 represents the compressive thrust in the arch.

Let $w(x,t)$ denote the motion about $y_s(x)$, i.e.,

$$w(x,t) = y(x,t) - y_s(x). \tag{10}$$

Then subtracting (8) from (5) yields the following equation of motion for w:

$$\ddot{w} + w'''' + \gamma^2 w'' - \psi(w'' + y_s'') = -\pi^4 p \tag{11}$$

where

$$\psi = 2\int_0^1 [(w')^2 + 2w'y_s']dx. \tag{12}$$

The boundary conditions on w are given by

$$w = 0,\ w'' = 0 \text{ at } x = 0,1. \tag{13}$$

If the ends are clamped, rather than pinned, the boundary conditions would be $w = 0$, $w' = 0$.

3. EQUILIBRIUM STATES. The equilibrium configurations $y_s(x)$ under the load $q(x)$ are determined from (8) and the boundary conditions. Treating γ^2 as a constant, (8) is solved for $y_s(x)$. The resulting expression is then used in (9) to yield the relation between the applied load q and the thrust γ^2 [4].

If the initial configuration $y_o(x)$ and the static load $q(x)$ acting on the pinned arch are

$$y_o(x) = \lambda \sin \pi x, \qquad q(x) = q_o \sin \pi x, \tag{14}$$

respectively, the primary equilibrium shapes have the form

$$y_s(x) = b \sin \pi x \tag{15}$$

where b is a real root of the cubic equation

$$b^3 + (1 - \lambda^2)\, b + q_o - \lambda = 0. \tag{16}$$

Other solutions, when real, are given by

$$y_s(x) = e_1 \sin \pi x + e_m \sin m\pi x, \qquad m = 2, 3, \ldots, \tag{17}$$

where

$$e_1 = \frac{q_o - \lambda}{m^2 - 1}, \quad e_m = \pm (\lambda^2 - m^2 - e_1^2)^{1/2}. \tag{18}$$

In a load-deflection space, solutions (17) bifurcate from the primary equilibrium path.

4. STABILITY OF EQUILIBRIUM. Again, assume $p(x,t) = 0$. The stability of the equilibrium configuration $y_s(x)$ can be investigated with the use of Liapunov's direct method. The state of the system is defined by the vector $u = (w,\dot{w})^T$, and the equilibrium configuration under investigation corresponds to $u = 0$. Define

$$V(u) = \int_0^1 [(\dot{w})^2 + (w'')^2 - \gamma^2 (w')^2]\, dx + (\psi/2)^2 \tag{19}$$

which corresponds to twice the total energy in the system, measured with respect to the state $u = 0$. Let k^2 be the dimensionless critical load for a column with the same boundary conditions as the arch (i.e., $k^2 = \pi^2$ if both ends are pinned, $k^2 = 2.046\pi^2$ if one end is clamped and the other is pinned, and $k^2 = 4\pi^2$ if both ends are clamped). Then

$$\int_0^1 (w'')^2\, dx \geq k^2 \int_0^1 (w')^2\, dx \tag{20}$$

for the appropriate class of admissible deflection functions [5]. Also, let K^2 be such that

$$\int_0^1 (w'')^2\, dx + (\psi/2)^2 \geq K^2 \int_0^1 (w')^2\, dx \tag{21}$$

for all admissible deflections w.

The use of the functional V in Liapunov's direct method leads to the following sufficient stability condition [6]: if

$$\int_0^1 (y_s')^2\,dx > \int_0^1 (y_o')^2\,dx - \frac{1}{2}K^2, \tag{22}$$

then the equilibrium state $u = 0$ is stable with respect to the metric

$$\rho = \left\{\int_0^1 [(\dot{w})^2 + (w'')^2]\,dx\right\}^{1/2} \tag{23}$$

or any metric equivalent to ρ. A weaker condition is obtained if K^2 is replaced by k^2 in (22). A condition independent of y_s is then given by

$$\int_0^1 (y_o')^2\,dx \le \frac{1}{2}k^2. \tag{24}$$

In other words, if the unloaded configuration $y_o(x)$ is sufficiently shallow so that (24) is satisfied, then the equilibrium shape under any static load will be stable with respect to ρ.

In order to obtain sufficient conditions for instability, consider the linearized form of (11), with $p = 0$:

$$\ddot{w} + w'''' + \gamma^2 w'' - 4\,y_s'' \int_0^1 w' y_s'\,dx = 0. \tag{25}$$

Define

$$W(u) = \begin{cases} -U(u)g(u) & \text{if } U(u) < 0 \\ 0 & \text{if } U(u) \ge 0 \end{cases} \tag{26}$$

where

$$U(u) = \int_0^1 [(\dot{w})^2 + (w'')^2 - \gamma^2 (w')^2]\,dx + 4\left[\int_0^1 w' y_s'\,dx\right]^2 \tag{27}$$

and

$$g(u) = \int_0^1 w\,\dot{w}\,dx. \tag{28}$$

Application of Liapunov's direct method with W furnishes the following result [6]: if $U < 0$ for some admissible deflection w and for $\dot{w} = 0$, then the equilibrium state $u = 0$ is unstable with respect to ρ. If w is chosen as the buckling mode w_1 of the corresponding column, this condition becomes

$$\int_0^1 (y_o')^2\,dx > \int_0^1 (y_s')^2\,dx + \frac{1}{2}k^2 + 2\,\frac{\left[\int_0^1 w_1'\,y_s'\,dx\right]^2}{\int_0^1 (w_1')^2\,dx}. \tag{29}$$

With the use of the Schwarz inequality, one can obtain the weaker instability condition

$$\int_0^1 (y_o')^2\,dx > 3\int_0^1 (y_s')^2\,dx + \frac{1}{2}k^2. \tag{30}$$

As an example, consider $y_o(x)$ and $q(x)$ given by (14). With $k^2 = \pi^2$, (24) yields the sufficient condition $\lambda^2 < 1$ for stability of the equilibrium configuration (15), and (22) furnishes the condition $\lambda^2 < 1 + b^2$ if K^2 is replaced by k^2. In order to determine the strongest stability condition, substitute

$$w = \sum_{n=1}^{\infty} a_n \sin n\pi x \tag{31}$$

into (21) and drop cubic and quartic terms in a_n. The resulting inequality is satisfied if one chooses

$$K^2 = \min\{\pi^2(1 + 2b^2),\ 4\pi^2\}, \tag{32}$$

and (22) yields

$$\lambda^2 < \begin{cases} 1 + 3\,b^2 & \text{if } b^2 \le 1.5 \qquad (33a) \\ 4 + b^2 & \text{if } b^2 > 1.5 \qquad (33b) \end{cases}$$

In terms of the initial arch rise λ and the loading parameter q_o, (33) furnishes the following stability conditions for the equilibrium shapes (15):

$$0 < \lambda < 1: \quad q_o < \infty\,; \tag{34a}$$

$$1 \le \lambda < \sqrt{5.5}\,, \quad b > 0: \quad q_o < \lambda + 2[(\lambda^2 - 1)/3]^{3/2}; \tag{34b}$$

$$\lambda \geq \sqrt{5.5}\,,\quad b > 0:\qquad q_o < \lambda + 3(\lambda^2 - 4)^{1/2}\ ; \tag{34c}$$

$$1 \leq \lambda < \sqrt{5.5}\,,\quad b < 0:\qquad q_o > \lambda - 2[(\lambda^2 - 1)/3]^{3/2}; \tag{34d}$$

$$\lambda \geq \sqrt{5.5}\,,\quad b < 0:\qquad q_o > \lambda - 3(\lambda^2 - 4)^{1/2}\ . \tag{34e}$$

For instability of the shapes (15), the inequality (30) becomes $\lambda^2 > 1 + 3\,b^2$. Therefore (33a) is necessary as well as sufficient for stability. The same can be shown for (33b) by choosing $w = \sin 2\pi x$, $\dot{w} = 0$ in the condition $U < 0$. The equilibrium configurations (17) are always unstable. In order to prove this, choose $w = \sin 2\pi x$ in $U < 0$ for the cases $m \geq 3$, and

$$w = 4\,e_2 \sin \pi x - e_1 \sin 2\pi x \tag{35}$$

for $m = 2$, with $\dot{w} = 0$.

If viscous damping is present, one can modify the functionals and prove that the shapes (15) are asymptotically stable if (34) is satisfied and unstable otherwise [7]. Domains of attraction also can be determined [8].

The behavior in the vicinity of a critical point can be investigated with the use of a perturbation technique. In particular, this type of analysis can be utilized to examine the effects of imperfections on the critical load [9]. This will be illustrated in Section 7.

5. VIBRATIONS. Small vibrations about an equilibrium configuration $y_s(x)$ are considered in this section. In the linearized equation of motion (25), let

$$w(x,t) = z(x)e^{i\omega t}, \tag{36}$$

where $z(x)$ denotes a vibration mode and ω is the corresponding frequency. Then (25) becomes

$$z'''' + \gamma^2 z'' - \omega^2 z - \alpha y_s'' = 0 \tag{37}$$

where

$$\alpha = 4 \int_0^1 z' y_s'\, dx. \tag{38}$$

One can solve for the modes and frequencies by first treating α as a constant in (37) and then substituting the resulting function $z(x)$ into (38). Since α is proportional to the amount of additional extension or contraction of the arch centerline during vibration, a mode z shall be called "inextensible" if $\alpha = 0$ and "extensible" if $\alpha \neq 0$.

Consider a pinned arch with symmetric initial shape $y_o(x)$ and a symmetric load $q(x)$. There will be a set of symmetric modes and a set of anti-symmetric modes. For vibrations about a symmetric configuration $y_s(x)$, the anti-symmetric modes will be inextensible and will be given by

$$z_n(x) = A_n \sin n\pi x, \quad \omega_n^2 = n^2\pi^2(n^2\pi^2 - \gamma^2), \quad n \text{ even}, \tag{39}$$

while the symmetric modes may be extensible or inextensible. In the special case of (14) and (15), the modes $z_n(x)$ are given by (39) for all values of n, and the frequencies are

$$\omega_1^2 = \pi^4(1 + 3b^2 - \lambda^2), \quad \omega_n^2 = \pi^4 n^2(n^2 + b^2 - \lambda^2), \quad n = 2,3,\ldots \tag{40}$$

This case is unusual in that the mode shapes do not depend on the load magnitude, and only $z_1(x)$ is extensible. Load-frequency curves for this case and several others are presented in Refs. [10,11]. The effects of an initial thrust and an elastic foundation are investigated in Ref. [12].

Rehfield [13] analyzed the nonlinear oscillations of a pinned, sinusoidal arch with no load. The frequency-amplitude relation is hardening for sufficiently small arch rise and softening otherwise. Reiss [14] considered the sinusoidal initial shape and load given by (14) and sinusoidal initial conditions. The nonlinear oscillations of the resulting single-mode response were examined with q_o at its critical value.

6. DYNAMIC LOADS. Assume the arch is subjected to a dynamic load $p(x,t)$ rather than a static load $q(x)$, in dimensionless terms. The critical load magnitude is often determined by the Budiansky-Roth criterion [15]. At increasing values of the load magnitude, a measure d of the maximum response is computed. The critical load is defined when, for a small increase in load magnitude, a large increase occurs in d. For very shallow arches, or for certain load distributions, such a jump in the maximum response may not occur, and the criterion may not be applicable.

In some cases, sufficient conditions for stability against dynamic snap-through may be obtained from energy considerations [3]. If all the equilibrium states can be determined, then the minimum amount of energy required for possible snap-through may be related to the load magnitude, for certain types of loads (e.g., impulse loads, step loads). Consider a pinned arch with $y_o(x)$ given by (14) and with $b > \sqrt{4.5}$, for example. If the initial velocity imparted by an impulse load to the arch satisfies the inequality

$$\int_0^1 (\dot{w})^2 \, dx < 4\pi^4(2b^2 - 3)/3, \tag{41}$$

then the arch will not exhibit snap-through [3].

An upper bound on the response can be derived by a Liapunov type of approach [16]. Let $q(x) = 0$, so $y_s(x) = y_o(x)$ and $\gamma^2 = 0$ in (19). It can be shown that

$$\dot{V}(t) \leq 2\, \beta(t) \sqrt{V(t)} \tag{42}$$

during motion, where

$$\beta(t) = \pi^4 \left[\int_0^1 p^2(x,t)\, dx \right]^{1/2}. \tag{43}$$

Therefore

$$\sqrt{V(t)} \leq \int_0^t \beta(\tau)\, d\tau. \tag{44}$$

One can then apply the inequalities

$$\int_0^1 (w'')^2 \, dx \geq c^2 \int_0^1 w^2 \, dx, \quad \int_0^1 (w'')^2 \, dx \geq h^2 w^2(x), \tag{45}$$

where $c^2 = \pi^2$, $1.56\pi^2$, and $2.07\pi^2$ and $h^2 = 48$, 101.9, and 192 for pinned-pinned, clamped-pinned, and clamped-clamped boundary conditions, respectively, to obtain the bounds

$$\int_0^1 w^2 \, dx \leq \frac{1}{c^2} \left[\int_0^t \beta(\tau) \, d\tau \right]^2 \tag{46}$$

and

$$|w(x)| \leq \frac{1}{h} \int_0^t \beta(\tau) \, d\tau, \qquad 0 < x < 1. \tag{47}$$

In order to obtain numerical solutions for the response, one can apply Galerkin's method to (11). For pinned ends, let w be given by (31) with $a_n = a_n(t)$ and $n = 1, \ldots, N$. For the case of (14) and (15), the equations for these amplitudes are found to be

$$\ddot{a}_n + \pi^4 n^2 (n^2 + b^2 - \lambda^2 + \phi) a_n + \pi^4 \, b\phi \, \delta_{1n}$$
$$= - 2\pi^4 \int_0^1 p \sin n\pi x \, dx \tag{48}$$

where

$$\phi = 2ba_1 + \sum_{n=1}^{N} n^2 \, a_n^2 \tag{49}$$

and δ_{1n} is the Kronecker delta. Numerical integration techniques can be utilized to determine $a_n(t)$, and the Budiansky-Roth criterion can be used to determine critical values of dynamic loads.

7. INFLUENCE OF LOAD POSITION. In this section, consider a single concentrated load at an arbitrary position $x = a$ along the arch. First, for quasi-static loading, let $p(x,t) = 0$ and

$$q(x) = q_a \, \delta(x - a), \tag{50}$$

where δ is the Dirac delta function. If the initial shape $y_o(x)$ is given by (14) and the ends are pinned, the solution of (7) and (8) yields the equilibrium configuration

$$y_a(x) = \pi^2 \lambda (\pi^2 - \lambda^2)^{-1} \sin \pi x + \pi^4 \gamma^{-3} \, q_a g(x) \tag{51}$$

where

$$g(x) = \{\sin[(x - a)\gamma] - (x - a)\gamma\} H(x - a) + (1 - a)\gamma x$$
$$- (\sin\gamma)^{-1} \sin[(1 - a)\gamma] \sin \gamma x , \qquad (52)$$

H is the Heaviside step function, and $\sin\gamma \neq 0$. The load-thrust relation is obtained by substituting (51) into (9). Snap-through instability occurs either at a limit point, where $dq_a/d(\gamma^2) = 0$, or at a bifurcation point, corresponding to $\gamma = 2\pi$. Here, bifurcation only governs if $a = 0.5$ and $\lambda > \sqrt{5.5}$ [2].

Critical values of q_a are plotted as dashed lines in Fig. 2, for values of the arch rise parameter $\lambda = 1.5$, 3, and 5. When $\lambda = 1.5$, the worst load position (corresponding to minimum critical load) is at the center of the span. When $\lambda = 3$ and $\lambda = 5$, the worst positions are at $a = 0.35$, 0.65 and $a = 0.33$, 0.67, respectively. The sharp drop in critical load in the neighborhood of the bifurcation load at $a = 0.5$ can be investigated by means of asymptotic expansions.

Assume the load position is $x = a - \varepsilon$, the corresponding equilibrium configuration is $y_s(x)$, and the critical load is q_c, with $y_s = y_a$ and $q_c = q_a$ when $\varepsilon = 0$. Let r be a perturbation parameter, and let

$$q_c = q_a + r\, q_1 + \frac{1}{2!} r^2 q_2 + \ldots , \qquad (53a)$$

$$y_s(x) = y_a(x) + r\, y_1(x) + \frac{1}{2!} r^2 y_2(x) + \ldots , \qquad (53b)$$

$$\delta(x - a + \varepsilon) = \delta(x - a) + \varepsilon\, \delta'(x - a) + \ldots , \qquad (53c)$$

$$\varepsilon = \varepsilon_1 r + \frac{1}{2!} \varepsilon_2 r^2 + \ldots . \qquad (53d)$$

These expansions are substituted into (8), with $q(x) = q_c\, \delta(x - a + \varepsilon)$, and coefficients of like powers of r are equated to zero. The coefficients of r^0 lead to (51). The coefficients of r^1 yield the equation

$$y_1'''' + \gamma^2 y_1'' - 4 y_a'' \int_0^1 y_a' y_1' \, dx = - \pi^4 [q_1\, \delta(x - a)$$
$$+ \varepsilon_1 q_a\, \delta'(x - a)], \qquad (54)$$

with boundary conditions (13) on y_1. Let z(x) denote the solution of the homogeneous part of (54). If one multiplies (54) by z, integrates over x, and integrates the term involving δ' by parts, one obtains

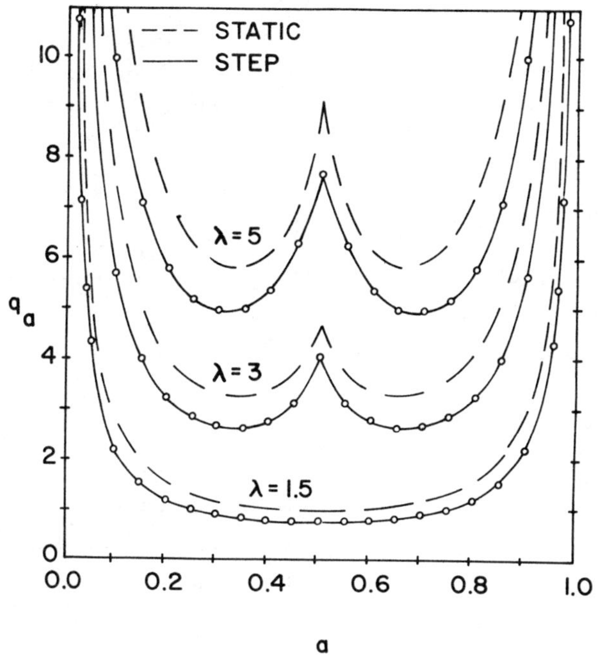

Fig. 2. Critical Static and Step Loads vs. Load Position

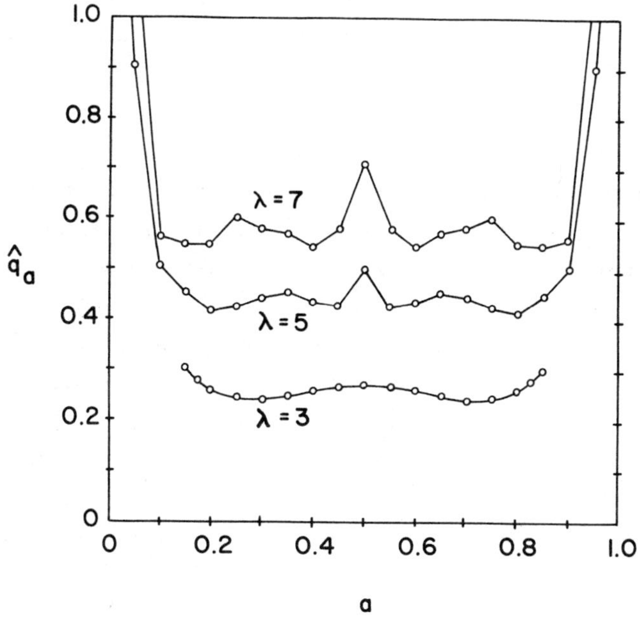

Fig. 3. Critical Impulse Loads vs. Load Position

$$q_1 z(a) - \varepsilon_1 q_a z'(a) = 0. \tag{55}$$

At the bifurcation points, $a = 0.5$ and $\gamma = 2\pi$, which lead to $z(x) = B \times \sin 2\pi x$. Then (55) furnishes $\varepsilon_1 = 0$, and (54) can be solved for $y_1(x)$. If the procedure is continued, one finds that [9]

$$q_c - q_a \sim \varepsilon^{2/3}, \tag{56}$$

i.e., the drop in critical load from the bifurcation point is proportional to the two-thirds power of ε.

Next assume a concentrated step load is applied to the arch, with

$$q(x) = 0, \quad p(x,t) = q_a H(t) \delta(x - a). \tag{57}$$

Numerical solution of (48) with $N = 5$ leads to the solid curves in Fig. 2 for $\lambda = 1.5$, 3, and 5 [17]. The curves are similar in shape to those for static loading, and have slightly lower values. Results for impulse loading, with

$$q(x) = 0, \quad p(x,t) = \hat{q}_a \delta(t) \delta(x - a), \tag{58}$$

are presented in Fig. 3 for $\lambda = 3$, 5, and 7. No sudden jump occurs in the response if the load is sufficiently close to the ends. For example, in the case $\lambda = 3$ in Fig. 3, critical loads are not defined for $x < 0.15$ and $x > 0.85$.

8. MULTIPLE LOADS. The interactive behavior of multiple, independent loads is often of interest. As an example, consider concentrated loads with magnitudes q_A, q_B at $x = 1/3$, $2/3$, respectively [18,19]. Interaction curves of critical load combinations (i.e., stability boundaries) are shown in Fig. 4 for $\lambda = 1.5$ and Fig. 5 for $\lambda = 3$, with the loads applied either quasi-statically (dashed curves) or stepwise in time (dark solid curves). For $\lambda = 1.5$, all critical static loads correspond to limit points. For $\lambda = 3$, a bifurcation point occurs when $q_A = q_B$, and the stability boundaries exhibit the sensitivity of this critical load to asymmetric imperfections. The other curves depicted in Figs. 4 and 5 correspond to constant values of the maximum response measure d defined by

$$d = \max_t \left\{ \int_0^1 w^2(x,t)\, dx \right\}^{1/2}, \tag{59}$$

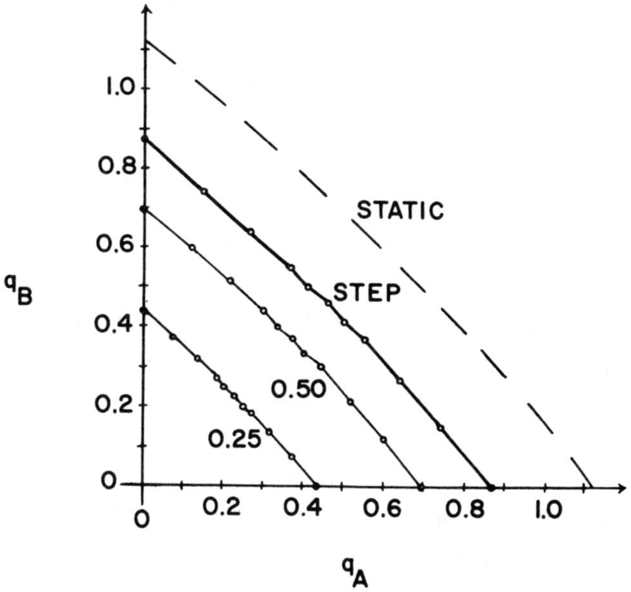

Fig. 4. Interaction Curves for Static and Step Loads; $\lambda = 1.5$

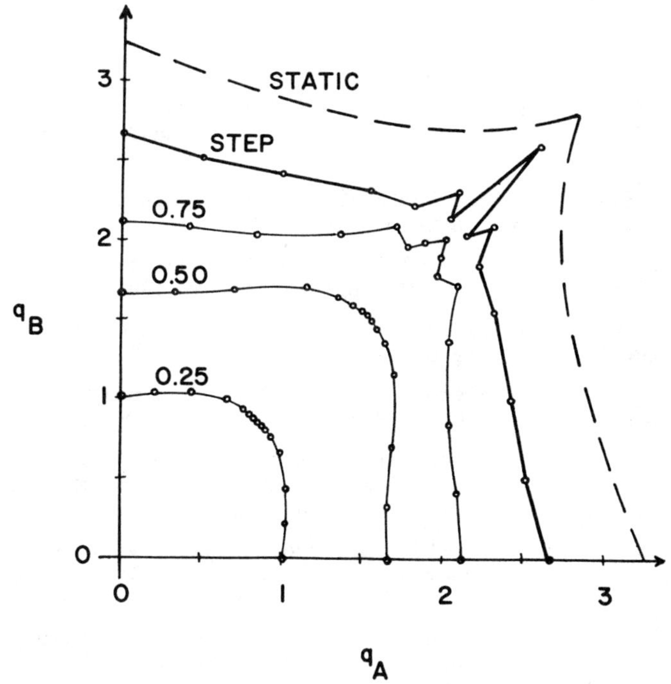

Fig. 5. Interaction Curves for Static and Step Loads; $\lambda = 3$

with the values of d listed on the curves.

9. CONCLUDING REMARKS. The behavior of shallow elastic arches has been studied by numerous researchers in structural mechanics. The governing equation of motion (or equilibrium equation in a static analysis) is nonlinear, but the nonlinearity is of a simple form. Equilibrium configurations can be determined analytically. The arches considered here exhibit deformations as soon as a load is applied, and become unstable by snap-through. These features are typical of shallow elastic structures, such as cylindrical panels and spherical caps, and the results for arches furnish insight into the behavior of these other structures (which are usually more difficult to analyze).

Only a few problems have been mentioned in this paper. Pinned, sinusoidal arches with sinusoidal loading distribution have been analyzed for the most part, because of the simplicity of the results. Many other cases have been treated in the literature. Static loads, step loads, and impulse loads have been considered. Other types of loads of interest include stochastic and harmonic loads (e.g., see Refs. [20,21]). Many interesting results on the vibrations and stability of shallow arches have been found, and many more will undoubtedly be uncovered in the future.

BIBLIOGRAPHY

1. R. H. Plaut, "Buckling of shallow elastic structures," in New approaches to nonlinear problems in dynamics, P. J. Holmes, ed., SIAM, Philadelphia, 1980, 361-375.

2. Y. C. Fung and A. Kaplan, "Buckling of low arches or curved beams of small curvature," NACA TN 2840, 1952.

3. C. S. Hsu, "On dynamic stability of elastic bodies with prescribed initial conditions," Int. J. Engrg. Sci., 4 (1966), 1-21.

4. H. L. Schreyer and E. F. Masur, "Buckling of shallow arches," J. Engrg. Mechs. Div. ASCE, 92 (1966), 1-17.

5. G. H. Hardy, J. E. Littlewood and G. Pólya, Inequalities, 2nd ed., Cambridge University Press, Cambridge, 1952.

6. R. H. Plaut, "Liapunov stability of shallow arches," in Developments in theoretical and applied mechanics, Vol. 9, R. M. Hackett, ed., Vanderbilt University, Nashville, 1978, 415-425.

7. R. H. Plaut, "Asymptotic stability and instability criteria for some elastic systems by Liapunov's direct method," Qtly. Appl. Math., 29 (1972), 535-540.

8. R. H. Plaut, "A study of the dynamic stability of continuous elastic systems by Liapunov's direct method," Rept. No. AM-67-3, College of Engineering, University of California, Berkeley, 1967.

9. R. H. Plaut, "Postbuckling analysis of continuous, elastic systems under multiple loads. Part 2: applications," J. Appl. Mechs., 46 (1979), 398-403.

10. E. R. Johnson and R. H. Plaut, "Load-frequency relationships for shallow elastic arches," Tech. Rept. No. VPI-E-80-4, College of Engineering, Virginia Polytechnic Institute and State University, Blacksburg, 1980.

11. E. R. Johnson, R. H. Plaut and C. C. Deel, II, "Load-frequency relationships for shallow elastic structures," in Recent advances in structural dynamics, M. Petyt, ed., University of Southampton, Southampton, 1980, 611-620.

12. R. H. Plaut and E. R. Johnson, "The effects of initial thrust and elastic foundation on the vibration frequencies of a shallow arch," J. Sound Vib., 78 (1981), to appear.

13. L. W. Rehfield, "Nonlinear flexural oscillations of shallow arches," AIAA J., 12 (1974), 91-93.

14. E. L. Reiss, "A new asymptotic method for jump phenomena," SIAM J. Appl. Math., 39 (1980), 440-455.

15. B. Budiansky and R. S. Roth, "Axisymmetric dynamic buckling of clamped shallow spherical shells," NASA TN D-1510, 1962, 597-606.

16. R. H. Plaut and E. F. Infante, "Bounds on motions of some lumped and continuous dynamic systems," J. Appl. Mechs., 39 (1972), 251-256.

17. W. E. Gregory, Jr., and R. H. Plaut, "Dynamic stability boundaries for shallow arches," J. Engrg. Mechs. Div. ASCE, to appear.

18. J. R. Cassidy and R. H. Plaut, "Buckling of shallow arches under combined distributed and concentrated loads," Mechs. Res. Comm., 8 (1981), 41-45.

19. W. E. Gregory, Jr., "Stability of a shallow arch under multiple dynamic loads," M. S. Thesis, Department of Civil Engineering, Virginia Polytechnic Institute and State University, Blacksburg, 1980.

20. S. T. Ariaratnam and T. S. Sankar, "Dynamic snap-through of shallow arches under stochastic loads," AIAA J., 6 (1968), 798-802.

21. N. C. Huang, "Dynamic buckling of some elastic shallow structures subject to periodic loading with high frequency," Int. J. Solids Strucs., 8 (1972), 315-326.

DEPARTMENT OF CIVIL ENGINEERING
VIRGINIA POLYTECHNIC INSTITUTE AND STATE UNIVERSITY
BLACKSBURG, VIRGINIA 24061

ABCDEFGHIJ–AMS–8987654321